5G Verticals

5G Verticals

Customizing Applications, Technologies and Deployment Techniques

Edited by

Rath Vannithamby
Intel Corporation
Portland, OR
USA

Anthony C.K. Soong
Futurewei Technologies
Plano, TX
USA

This edition first published 2020
© 2020 John Wiley & Sons Ltd

The right of Rath Vannithamby and Anthony C.K. Soong to be identified as the authors of the editorial material in this work has been asserted in accordance with law.

Registered Offices
John Wiley & Sons, Inc., 111 River Street, Hoboken, NJ 07030, USA
John Wiley & Sons Ltd, The Atrium, Southern Gate, Chichester, West Sussex, PO19 8SQ, UK

Editorial Office
The Atrium, Southern Gate, Chichester, West Sussex, PO19 8SQ, UK

For details of our global editorial offices, customer services, and more information about Wiley products visit us at www.wiley.com.

Wiley also publishes its books in a variety of electronic formats and by print-on-demand. Some content that appears in standard print versions of this book may not be available in other formats.

Library of Congress Cataloging-in-Publication Data
Names: Vannithamby, Rath, editor. | Soong, Anthony C.K., editor.
Title: 5G verticals : customizing applications, technologies and deployment
    techniques / edited by Rath Vannithamby, Anthony C.K. Soong.
Description: First edition. | Hoboken, NJ : John Wiley & Sons, Inc., 2020.
    | Series: IEEE series | Includes bibliographical references and index.
Identifiers: LCCN 2019045542 (print) | LCCN 2019045543 (ebook) | ISBN
    9781119514817 (hardback) | ISBN 9781119514831 (adobe pdf) | ISBN
    9781119514855 (epub)
Subjects: LCSH: 5G mobile communication systems.
Classification: LCC TK5103.25. A175 2020 (print) | LCC TK5103.25 (ebook)
    | DDC 621.3845/6–dc23
LC record available at https://lccn.loc.gov/2019045542
LC ebook record available at https://lccn.loc.gov/2019045543

Cover Design: Wiley
Cover Image: © spainter_vfx/Shutterstock

Set in 9.5/12.5pt STIXTwoText by SPi Global, Pondicherry, India

Printed and bound by CPI Group (UK) Ltd, Croydon, CR0 4YY

10   9   8   7   6   5   4   3   2   1

Contents

List of Contributors

S.M. Ahsan Kazmi
Innopolis University, Innopolis, Russia

Sergey Andreev
Tampere University, Tampere, Finland

Jingwen Bai
Intel Corporation, Santa Clara, CA, USA

Mehdi Bennis
University of Oulu, Oulu, Finland

Tri Nguyen Dang
Kyung Hee University, Seoul, Korea

Margarita Gapeyenko
Tampere University, Tampere, Finland

Amitava Ghosh
Nokia Bell Labs, Naperville, IL, USA

Choong Seon Hong
Kyung Hee University, Seoul, Korea

John Kaippallimalil
Futurewei Technologies, Plano, TX, USA

Yevgeni Koucheryavy
Tampere University, Tampere, Finland

Ulas C. Kozat
Futurewei Technologies, Santa Clara, CA, USA

Christian Maciocco
Intel Corporation, Hillsboro, OR, USA

Dmitri Moltchanov
Tampere University, Tampere, Finland

Murali Narasimha
Futurewei Technologies, Rolling Meadows, IL, USA

Vitaly Petrov
Tampere University, Tampere, Finland

Ana Lucia Pinheiro
Intel Corporation, Hillsboro, OR, USA

Rapeepat Ratasuk
Nokia Bell Labs, Naperville, IL, USA

Tony Saboorian
Futurewei Technologies, Plano, TX, USA

Andrey Samuylov
Tampere University, Tampere, Finland

Anthony C.K. Soong
Futurewei Technologies, Plano, TX,
USA

M. Oğuz Sunay
Open Networking Foundation, Menlo
Park, CA, USA

Shilpa Talwar
Intel Corporation, Santa Clara, CA, USA

Nguyen H. Tran
The University of Sydney, Sydney,
NSW, Australia

Rath Vannithamby
Intel Corporation, Portland, OR, USA

Frederick Vook
Nokia Bell Labs, Naperville, IL,
USA

Amanda Xiang
Futurewei Technologies, Plano, TX,
USA

Feng Xue
Intel Corporation, Santa Clara, CA,
USA

Jin Yang
Verizon Communications Inc.,
Walnut Creek, CA, USA

Shu-ping Yeh
Intel Corporation, Santa Clara, CA,
USA

Preface

Looking back at the technical history of wireless communication systems, it becomes apparent that the fifth generation (5G) is following a fundamentally different developmental tract than what came before it. The first generation was based upon analog transmission technology. The second generation was the transition from analog to digital transmission technology in the form of mainly time division multiple access. The third generation was the transition of the wireless industry to the Code Division Multiple Access (CDMA) wave form. The fourth generation was the adoption of Orthogonal Frequency Division Multiplexing (OFDM). Each generation was associated with a change in the waveform and an associated improvement in the spectral efficiency due to the waveform. However, 5G is decidedly not a change in the waveform because OFDM is still the waveform in the first release of 5G; Release 15. The main reason for this, in our opinion, is that in the frequency range from 1 to 60 GHz, OFDM is extremely flexible and a proven waveform does not exist in the literature that significantly outperforms it.

This raises the question: What is so special about 5G leading to all the hype? In the literature and in the press, there are numerous articles about how 5G will be "more"; more capacity, more coverage, higher data rates, and lower latency. Yes, 5G is about all of that but in our opinion, these aspects are insignificant to the revolution that is 5G. To understand its significance, we must look at the environment onto which 5G will be deployed.

As humans we have gone through approximately three industrial revolutions. The first industrial revolution was the harnessing of steam power for manufacturing. Humans experienced more interaction and communication sped up with the invention of the steam railroad. The second industrial revolution was the harmonization of science in manufacturing processes which resulted in mass production and the assembly line. Associated with the industrial advances was another order of communication improvements with the development of the telegraph and telephony. The third industrial revolution witnessed the rise of electronics and ushered in the digital age that we are enjoying today. It started the era of miniaturization which notably opened up new space and biotechnical frontiers. Communication was again revolutionized with the development of untethered wireless communication. What is clear is that associated with every industrial revolution there was an improvement in human communications by orders of magnitude. Communication and industrial revolution can then be considered different sides of the same coin.

5G is arriving as the fourth industrial revolution is unfolding. This industrial revolution is about the merging of the industrial physical world with the biological and the digital worlds. One characteristic which is already clear is that digitalization and artificial intelligence has evolved to a point where we can build a new virtual world from which we can control the physical world. In the communication industry literature, the Internet of Things (IoT) or vertical industry evolution fall under the umbrella of enabling the fourth industrial revolution by 5G. Unfortunately, due to its diversity, the information about how 5G will support this industrial revolution is scattered among a number of papers, keynote presentations, and panel discussions. This motivated us to try to gather and harmonize the different bits of information into one book.

This book is written by experts in the field and experts who developed the 5G standards. It not only discusses technically the 5G system as it is standardized in Release 15 but it also discusses the technology that enables the 5G system to be flexibly deployed and scaled as new use cases that support the vertical industry are identified and developed. It also looks at the market forces that are driving how 5G will support the vertical use cases. Furthermore, several specific vertical industries, which have the potential to be among the first industries to use 5G, will be analyzed in detail. It will demonstrate that 5G together with the fourth industrial revolution has a chance to break the link between development and environmental cost. For the first time, the efficiencies enabled by 5G may allow for economic development without sacrificing the environment.

Portland, OR and Plano, TX, June 2019   *RV and ACKS*

Acknowledgments

This book would not have been possible without the support of others. We would like to gratefully acknowledge our colleagues who have contributed chapters. Not only have they shaped the commercial wireless industry but, over many years, we have also learned a great deal from them. We would also like to thank our families for their encouragement, support, and undying love. Special thanks go to our wives, Srirajani and Sherry, who have endured our absence for many days while we were away at standards and industrial meetings developing the 5G standards. A special acknowledgment goes to the editorial team and production team at John Wiley; especially to our project editor, Louis Manoharan.

Part I

Introduction to 5G Verticals

1

Introduction

Anthony C.K. Soong[1] and Rath Vannithamby[2]

[1] *Futurewei Technologies, Plano, TX, USA*
[2] *Intel Corporation, Portland, OR, USA*

Abstract

This chapter looks at the features in the first phase of fifth generation (5G) that supports the vertical industry. Moving onto future phases of 5G, Third Generation Partnership Project and the cellular industry has invited and offered to work with all vertical industries to define additional requirements for future 5G releases. One interesting vertical industry, for example, is professional audio production which requires strict synchronization of devices to function. The chapter discusses the significance of network slicing for the support of the vertical industry services. It examines a number of other issues that impact the vertical services such as virtual network function placement, multi-access edge computing, and artificial intelligence. The chapter further presents an overview of the key concepts discussed in the subsequent chapters of this book. The book examines the impact of 5G cellular communications on the various vertical industries.

Keywords *5G cellular communications; network issues; network slicing; radio access; Third Generation Partnership Project; ultra-reliable and low-latency communications standardization; vertical industries*

1.1 Introduction

It is without a doubt that commercial wireless cellular communication has changed how we, as human beings, interact with each other. As young graduate students over 30 years ago, we could not have imagined how enhancing our

5G Verticals: Customizing Applications, Technologies and Deployment Techniques, First Edition.
Edited by Rath Vannithamby and Anthony C.K. Soong.
© 2020 John Wiley & Sons Ltd. Published 2020 by John Wiley & Sons Ltd.

communications would have altered the way we interact in the world. At that time, our dream was to bring ubiquitous wireless telephony to the world. The attraction of that dream is obvious, to give users the freedom of movement while being able to continue a telephone call. There was no discussion about the killer application because voice was the only application. Commercial wireless systems were in their infancy, analog modulation was king, and only the very elite would have a telephone in their car. If you wanted to take the phone with you, it was carried in its own briefcase because the size of the phone was that large.

Decades ago, the concept of cellular was considered new. It necessitated the development of base stations serving a small area, or cell, because of the requirement of providing high capacity mobile telephony that did not require a very large number of channels. Arguably, cellular communication was born on 4 January 1979, when the Federal Communications Commission authorized Illinois Bell Telephone Co. to conduct a trial of a developmental cellular service in the Chicago area. Around the same time, American Radio Telephone Service Inc. was authorized to operate a cellular service in the Washington–Baltimore area. The feasibility and affordability of cellular services where the same channel may then be re-used within a relatively small distance was then demonstrated. More importantly, from a communication system point of view, the concept of capacity increase by densification[1] was established and full commercial service first began in Chicago in October 1983.

From these humble beginnings, cellular communication has grown into a common and necessary part of everyday human interaction. The transformation of the cellular phone from a telephony device to a pillar of human social interaction can be laid at the development of the so-called "smart phone". Humanity changed in 2008 and is now dependent upon the applications on the smart phone as a harbinger of information, as well as to enable socialization with others in an individualized way. Indeed it is ubiquitous customized socialization that has transformed our society. In some sense, it has not only brought us closer to each other but also closer to our humanity.

As we move into the era of 5G cellular communication, humanity will move from the age of human social communication to a world where communications is fusing together the physical, digital, and biological worlds; the so-called "fourth industrial revolution" [3]. This will represent an unprecedented opportunity to transform our industry and simultaneously drive profitability and sustainability. Innovations will enhance the production cycle and connect manufacturers with their supply chains and consumers to enable closed feedback loops to improve products, production processes, and time-to-markets. As an example of

---

1 It should be noted that almost all of the gain in the capacity of the commercial wireless communication system in history has come from densification of the network. A discussion of fifth generation (5G) densification can be found in [1, 2].

the potential economic impact, the World Economic Forum studied the impact of this on the state of Michigan [4]; the birthplace of modern automotive mass manufacturing. Today, the manufacturing environment of the automotive industry is undergoing unprecedented change spurred on by the changing expectation of digital consumers, the emergence of the smart factory, and the rise of connected vehicles.[2] This change is coinciding with the transition from mass manufacturing to hyper-customization demanded by the customer. No longer are consumers looking to just buy a product; they are looking for a complete customized user-centric experience.

Sustainability has now also become a prime concern for consumers and manufacturers alike. Embracing sustainability and green principles is not just a mere marketing tool anymore. Consumers and regulators are no longer satisfied when economic growth happens at the expense of the environment and are setting higher and higher sustainability requirements. The fourth industrial revolution presents an opportunity to decouple this relationship by providing both economic growth while enhancing simultaneously the environment [4]. As an example, the auto manufacturers are, more and more, making bold commitments to sustainability. The vision of General Motors (GM) of zero crashes, zero emissions, and zero congestion demonstrates the auto industry's strategy of coupling growth with sustainability. As part of this, GM is committed to using 100% renewable energy by 2050. Both GM [5] and Ford [6] have committed to the United Nations 2030 sustainability goals [7].

For the state of Michigan alone, the fourth industrial revolution is projected to add $7 billion to the automotive industry by 2022 [4]. From a global perspective, the main societal benefits are the time and cost savings and efficiency gains from specific hyper-customized services and applications while enhancing sustainability. Paramount within this transformation is that human-to-human communication will only form one pillar. The other pillars will be the revolutionary changes to other vertical industries enabled by ubiquitous communication. This book will examine the impact of 5G cellular communications on the various vertical industries.

1.2 5G and the Vertical Industries

As the fourth generation (4G) cellular system, embodied by Long-Term Evolution (LTE), is now reaching maturity, the cellular industry has developed the first standards for 5G: Third Generation Partnership Project (3GPP) Release 15 [8, 9]. 5G, however, arrives at a challenging time for the industry because the industry

---

2 See Chapters 9 and 10.

has reached 8.8 billion global mobile connections from 5.1 billion unique subscribers [10]. This implies that almost every person who wants a mobile connection already has one. The economic impact is such that increasing revenue for mobile carriers from enhanced mobile broadband (eMBB) service will be difficult. The first wave of 5G users will not be new users but mostly users that are upgrading their services from older generations to the 5G eMBB service. It can then be argued that the success of 5G, for the industry, will depend upon more than the success of the 5G eMBB service. In other words, one must look beyond the eMBB service in order to expand the 5G footprint.

There is, however, a pot of gold at the end of the rainbow. It is expected that in the future, the number of connected things will far exceed the number of humans on Earth (7.6 billion). One research report predicted that by 2020 the number of connected devices will reach 20 billion [11]; roughly three times the number of humans on Earth. Furthermore, the growth of this number is expected to be exponential and unbounded. Consequently, connected devices (the so-called Internet of Things [IoT]) provides a growth path for the cellular industry.

This growth path, however, is not without its challenges. While the eMBB service is a monolithic service, the IoT potential includes a large number of vertical industries; each with its own service requirements. Indeed the service is so diverse that it resemble the long tailed services discussed in [12] with eMBB as the head service and the vertical services representing the long tail (Figure 1.1). The main characteristic of the long tail is that each vertical service in the long tail does not generate significant revenue but because the tail is long, there is a large number of different vertical services, the total revenue is significant; often more than the so-called "head service". This is the classic so-called "selling less to more" scenario.

The size of the long tail has been examined by a number of studies. For example, [13] estimated that the economic impact would be a massive 2016 $12 trillion per year by 2035 (Figure 1.2). A similar projection of $14 trillion by 2030 was reported by [14]. To give a sense of the scale, the current global revenue of the cellular industry is about $1 trillion per year in 2018 [10]. The total size of the tail has the potential to be 12 times that of the head in 2018. Even capturing a part of this would represent significant growth for the cellular industry.

The European Commission has an in-depth study on the growth in vertical business for the mobile 5G service providers [15]. This report focused on just a few key verticals but the trend is very clear. It showed that the revenue mix in 5G, is perhaps, the biggest difference between 5G and 4G (Figure 1.3). Currently the revenue from IoT services is a very small part and most of the revenue is from the eMBB service. By 2025, however, it is anticipated that the revenue from the vertical services would surpass that of eMBB. This would be just as transformative a milestone in the history of the commercial wireless industry as the first time that

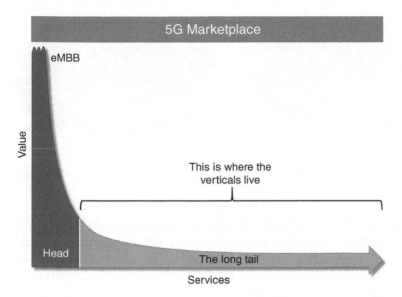

Figure 1.1 A schematic representation of the vertical services as long tailed services. The figure Is modified from that in [12] to reflect the wireless industry.

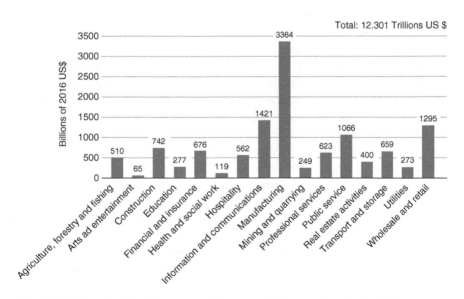

Figure 1.2 An estimate of the economic impact of different industries by 2035. *Source:* Data from [13].

CURRENT   2025

Figure 1.3 The projected operator revenue from difference services. *Source:* Projection data from [15].

the revenue from data exceeded that from voice. 5G has the capability for such a transformation because it offers more to the vertical industry than just connectivity. It offers an innovative communication platform to not only do things more efficiently but to do more and different kind of things; ranging from smart factories to smart cars to smart unmanned aerial vehicles.

From service providers' point of view, the road to 5G is clear; they have to begin to restructure their businesses around the vertical industry opportunities that 5G provides. Several tier one operators have already begun to restructure their organization to take advantage of these new opportunities. In the past, the operator might have been divided into a mobile, an enterprise, and maybe an IoT business unit. Today, we are starting to see them structure by vertical industries. This allows them to provide the needed focus on the new business initiatives built around specific vertical industries.

While carriers are at the beginning of their journey to support the vertical industry, some of the ecosystem players are further ahead [14]. Intel, Qualcomm, and Huawei are well on their way to building interoperable integrated solutions across diverse vertical sectors. Apple and Samsung are providing horizontal interoperability between devices with their integrated home IoT suites. The commercial mobile industry will provide the key backbone that supplies essential connectivity between billions of cloud applications and sensors with the biggest opportunities in providing consumer and enterprise applications and services to the verticals making the IoT a reality.

Mobile carriers are not the only ones that can see this new opportunity. Over the top (OTT) players, such as Amazon that started the whole heavy tail transformation, are also extremely excited about the potential new 5G business opportunities. They are also transforming themselves to provide new customized services for the vertical market. The carriers, if they are to stop the trend, namely OTT provides the service and the carriers just provide the pipe for the communications that is needed by the services, that 4G started, will need to find ways to out innovate the OTT players. All is, however, not bleak for the carriers. As the rest this book will make evident, 5G is ushering in the age of ultra-customization. This means that the competitive advantage for the carriers is laid in the unique network intelligence that allows the carriers to customize their 5G service. For example, they can use machine learning techniques to customize the network slice for the application that will provide a better service experience.

Socioeconomically 5G support of the vertical industry has the potential to offer the most promise for eliminating the disadvantages associated with the digital divide [16]. The access to mobile broadband, from 4G, has significantly closed the digital divide by providing broadband access to those without fixed broadband access. The socioeconomic disadvantage, however, persisted even though access to broadband improved because the digital divide and physical divide are often in conjuction. The physical isolation of these vulnerable populations lock them out of opportunities and services even if they are provided with a broadband link. For example, they are unable to maintain regular contact with service providers which hampers their ability to monitor chronic diseases, connect with job opportunities real time, or connect with assistance for homework and research papers to aid their academic pursuits. The support of the different verticals with 5G can potentially solve all these issues by bringing the services to the disadvantaged. It can untether expert medical care from physical hospitals, expert career coaching from physical offices, and expert teachers from their academic institution, to offer customized services to their clients. The report [16] studied a number of vertical use cases and concluded that 5G vertical support, along with the hyper-customized technologies and applications that they will support, can be the great socioeconomic equalizer by providing emerging pathways for economic and social opportunities.

The concept of hyper-customization, selling less to more, has a profound impact on the requirement of the cellular systems. For one thing, it is no longer economical to build a customized system for every service. That does not mean that each vertical service does not need customization but just that such customization can no longer be at the hardware level. Fortunately, at about the same time that 5G was being developed, the phenomenon of movement of data to the cloud so that it can be easily accessed from anywhere with any device was taking shape [1]. The end points and the time frame for which network services are providing are thus fundamentally redefined. The resulting network needs to be much more nimble, flexible, and scalable. The two enabling technologies are: network function virtualization (NFV) and software defined networking (SDN). Network functions that have been traditionally tied to customized hardware appliance via NFV can now be run on a cloud computing infrastructure in the data center. It should be understood that the virtualized network function here does not necessarily have a one to one correlation with the traditional network functions but rather can be new functions created from, e.g. the functional decomposition of traditional functions. This gives the network designer extreme flexibility on how to design the virtualized functions and interconnection. SDN provides a framework for creating intelligent programmable networks from the virtualized network functions that are, possibly, interconnected on all levels [17]. Together these technologies endow the 5G network with significant nimbleness through the creation of virtual network slices that support new types of vertical services [18] that may leverage information centric network features [19]. A more precise and detailed discussion of network slicing will be given later, for now we will just use the concept that a network slice contains all the system resources that are needed to offer a particular services so that each vertical industry can be supported in a slice. Consequently, 5G solves the economic issue by affording the network designer to create a single network that can be customized (slice) and scaled [20] for different vertical markets economically via software.

1.3   5G Requirements in Support of Vertical Industries

There are now a significant number of papers in the literature that discuss 5G systems support for vertical industries. The main characteristics of these systems are rapidly becoming open ecosystems built on top of common infrastructures [21]. In essence, they are becoming holistic environments for technical and business innovation that integrates network, computer and storage resources into one unified software programmable infrastructure. Moreover, the strict latency requirements of verticals, such as factory 2.0, is forcing the network designer to put significant network resources at the edge of the network for distributed computing. This is the so-called multi-access edge computing (MEC) phenomenon.

The 5G service and operational requirements have been detailed in [22]. Detailed requirements from the automotive, eHeath, Energy, Media, and Entertainment, and Factory of the Future were analyzed. The needs of these industries can be condensed into five use cases: dense urban information society (UC1), virtual reality office (UC2), broadband access everywhere (UC3), massive distribution of sensors (UC4), and actuators and connected car (UC5). These can then map to the following radio network requirements:

1) 300 Mbps per user in very dense outdoor and indoor deployments.
2) Broadband access (50 Mbps per user) everywhere else.
3) 1 Gbps per user for indoor virtual reality office.
4) Less than 10 ms exchange to exchange (E2E) latency everywhere.
5) One million devices per square kilometer.

6) Battery life of 10 years.
7) Ultra reliable communication with 100 Mbps per user with 99.999% reliability.
8) Ultra low latency communication with 5 ms E2E latency.

These requirements were developed into the standards [9] as eMBB, massive machine type communication (mMTC) and ultra-reliable and low-latency communications (URLLC) services. Since the use of unlicensed carriers in coordinated, on demand service-orientated fashion can offer high performance system gains [23] for certain vertical industry services, the standards also developed the support for Licensed Assisted Access.

The role of the network operator can clearly been seen as to provide a tailored communication system to its customers (end users, enterprises, and vertical industries). This requires the 5G system to have the ability to flexibly integrate the cellular communication system for various business scales ranging from multi-nationals to local micro-businesses [24]. Having to adopt to a variety of requirements, some of which may not be known during initial system design, has led to the concept of network slicing. This network flexibility will, thus, become a first key design principle in the 5G control plane. Other key design principles which allows the customer to customize, manage and even control the communication slice include: the openness of the control plane for service creation, connectivity via a multitude of access technologies (e.g. licensed and unlicensed access) and context awareness by design. The technology enablers are:[3] network slicing, smart connectivity, modular architecture, resource awareness, context awareness, and control without ownership. Some of these concepts, such as control without ownership, are still under discussion in 3GPP and will not be finalized in the first release of the standards.

The aforementioned requirements are the first step in supporting the vertical industries. More detailed requirements for supporting the different verticals will be discussed in later chapters. The remainder of this chapter will look at the features in the first phase of 5G that supports the vertical industry. Moving onto future phases of 5G, 3GPP and the cellular industry has invited and offered to work with all vertical industries to define additional requirements for future 5G releases. One interesting vertical industry, for example, is professional audio production which requires strict synchronization of devices to function [25]. The requirement for low latency and synchronization goes beyond what is being developed in the current URLLC standardization. 3GPP will consider these requirements and the suite of options for vertical support in 5G in future releases of the standard will, thus, be enhanced as needed.

---

3 A more detailed discussions of the 5G network technologies can be found in Chapters 6 and 7.

1.4 Radio Access

The 3GPP has been working on specifying the 5G radio interface, also referred to as New Radio (NR) [26].[4] There is always a need for mobile broadband with higher system capacity, better coverage, and higher data rates. In 5G, the needs are for more than mobile broadband. One such need is for URLLC, providing data delivery with unprecedented reliability in combination with extremely low latency, for example targeting industrial applications in factory setting.

Another is for mMTC, providing connectivity for a very large number of devices with extreme coverage, very low device cost and energy consumption. NR is being designed to support eMBB, URLLC, and mMTC. Moreover, it is possible that other use cases that are not yet known may emerge during the lifetime of NR. Therefore, NR is designed to support forward compatibility, enabling smooth introduction of future use cases within the same framework.

At present, 3GPP plans to develop the technologies and features to support eMBB, URLLC, and mMTC over multiple releases, e.g. 3GPP Releases 15, 16, and 17. The Release 15 specification covers technologies and features that are needed to support eMBB and URLLC. 3GPP has developed technologies, e.g. enhanced machine type communication and narrowband IoT that can handle mMTC use cases and can satisfy mMTC 5G requirements, and further technologies are expected to be developed to support other mMTC use cases in the later releases.

The eMBB service is to support a range of use cases including the ones identified in [27], namely (i) broadband access in dense areas, (ii) broadband access everywhere, and (iii) higher user mobility. Broadband access must be available in densely populated areas, both indoors and outdoors, such as city centers and office buildings, or public venues such as stadiums or conference centers. As the population density indoors is expected to be higher than outdoors, this needs correspondingly higher capacity. Enhanced connectivity is also needed to provide broadband access everywhere with consistent user experience. Higher user mobility capability will enable mobile broadband services in moving vehicles including cars, buses, and trains.

The URLLC is a service category designed to meet delay-sensitivity services such as industrial automation, intelligent transportation, and remote health [28]. Since the human reaction time is in the order of a millisecond (e.g. around 1 ms

---

4 A more detailed discussion of how the industry developed the 5G standards, its technology and evaluated the performance can be found in [8].

for hand touch and 10 ms for visual reaction), packets for the mission-critical applications should be delivered in the order of a microsecond [28].

The NR is designed in such a way that eMBB, URLLC, and mMTC use cases are supported over a unified, flexible, and scalable frame structure. The NR interface provides a flexible framework that can be used to support different use cases. This is accomplished through a scalable Orthogonal Frequency Division Multiplexing (OFDM) based numerology and flexible frame structure. When scheduling URLLC traffic together with eMBB traffic within the same frame, since the URLLC traffic needs to be immediately transmitted due to its hard latency requirements, the URLLC transmission may overlap onto previously allocated eMBB transmissions. Various techniques to efficiently provide the resource allocation for eMBB and URLLC are given in [29, 30].

The service requirements of 5G also cause significant changes in the concept of a cellular system. For example, the concept of a cell is no longer relevant. It has evolved, in 5G, to a concept known as multi-connectivity. In principle, multi-connectivity refers to a device sharing resource of more than one base station. This concept is not really new and it has its roots in Release 12 dual connectivity. For 5G FDD/TDD dual connectivity, this allows for the transmission of multiple streams to a single UE that is semi-statically configured. The higher layer parallel transmission here is the key. It makes dual connectivity applicable to those deployment scenarios without requiring ideal, almost zero latency backhauls. 5G support LTE-NR dual connectivity which allows one of the carriers to be LTE while the other is NR.

Also within the multi-connectivity umbrella, it allows for uplink (UL)/downlink (DL) decoupling; having different cells associated with the UL and DL. The basic configuration is for one cell to be configured with two ULs and one DL. One of the UL carriers is a normal TDD or FDD UL carrier while the other is a supplementary uplink[5] (SUL) band (Figure 1.4). This configuration allows for the dynamic scheduling and carrier switching between the normal UL and the SUL. The UE is configured with two ULs for one DL of the same cell, and uplink transmissions on those two ULs are controlled by the network. One major advantage of the SUL is for it to be in a lower frequency while the paired UL and DL carrier is in a higher frequency. This is extremely important because the UL is power restricted for health reasons. For high data rates, configuring the SUL in this configuration can give extra range to the UL to balance the UL and DL coverage. This capability is key to certain vertical use cases that are more UL intensive.

---

5 In 5G, an UL-only carrier frequency is referred to as the SUL frequency from a NR perspective. See [8].

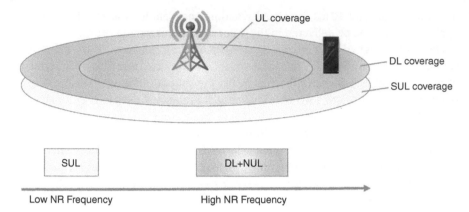

Figure 1.4 Single cell with two uplink carriers and one downlink carrier. NUL, normal uplink.

1.5 Network Slicing

It should now become clear to the reader that we cannot understate the importance of network slicing in supporting the vertical industries. On account of the recent advances in the NFV and SDN technologies [31], mobile network operators (MNOs) or even mobile virtual network operators[6] (MVNOs) can use network slicing to create services that are customized for the vertical industries [32].

The impact of slicing in a 5G system was studied in [33] for the specific case of a vehicle network. It confirmed that slice isolation is a key requirement in three different aspects: (i) slice A should not be influenced by other slices even when the other slices are running out of resources, (ii) to prevent eavesdropping, direct communication between the slices should not be allowed, and (iii) a mechanism to prevent "through the wall hijacking" should be in place. Security isolation was the topic of investigation in [34]. Their conclusion is that a prerequisite for running a highly sensitive service in a network slice is full isolation of the slice from all other users. This needs a trusted relation between the vertical industry and the mobile operator from which the vertical is renting the resources. The operator needs to ensure to the vertical that it will guarantee the needed isolation and security for their traffic and devices. Currently in 3GPP there is a study item to look at the isolation issue. One proposal that is favored by some of the industrial 2.0 companies is for the vertical industry to take full control over the network slice. This

---

6 The difference between a MNO and a MVNO is that the MNO owns the underlying network resources. A MVNO offers mobile services like a MNO but does not own the network resources. It rents the resources from one or more MNOs.

allows for the vertical industry to dictate the isolation and security. It should be noted that this point is still begin discussed in 3GPP and the solution is currently not yet settled.

The feasibility of using NFV and SDN for slicing has been recently studied. In [35] a network slicer which allows vertical industries to define vertical services based upon a set of service blueprints and arbitrating, in the case of resource shortages, among the various vertical industries was shown to be effective for network slicing implementation. In a similar concept, a slice optimizer which communicates with an SDN controller to receive information regarding the network slice and adapts the slices according to the network state was proposed and analyzed in [36].

In parallel to the NFV and SDN advances, semantic interoperability development allows for the exchange of data between applications, as well as an increase in the level of interoperability, analytics, and intelligence [37]. This technology is now overcoming the limitation of static data models and bridges the gap between the different vertical industries' network slices.

An overview of the solutions proposed in the literature [38, 39], and the current network slicing status in 3GPP are detailed in [40]. A network slice is defined, in the standards, as a logical network that provides specific network capabilities and network characteristics. From a user's point of view, it behaves like a customized wireless system, which may be dynamically reconfigured, dedicated for its use. The standards have taken a strong step toward a cloud native approach to network slicing [41]. This is achieved through virtualization and modularization of the network function while providing a notion of network programmability through a system service-based architecture. The intension is for network function services to be flexibly used by other authorized network functions by exhibiting their functionality via a standardized service-based interface in the control plane. The question as to whether the slice extends and how far into the radio access network (RAN) [42] was discussed extensively in 3GPP. RAN slicing is particularly challenging because of the inherently shared nature of the radio channel, the desire for multi-user diversity gain, and the impact of the transmitter on the receiver [43]. The conclusion is that for the first phase of 5G, it was agreed that the RAN will only be slice-aware so as to treat the sliced traffic accordingly. The RAN will support inter-slice resource management and intra-slice quality of service differentiation. The network designer has full flexibility of how to achieve this in the system.

It is now clear that NFV and SDN allow network resources to be isolated into a programmable set of slices in order to guarantee the E2E performance of the network. Note that this guarantee is regardless of what is happening elsewhere in the network. This is known as "slice isolation". We will see later on that slice isolation is one of the reasons that makes the design of the 5G network so challenging because one cannot just blindly implement the IT cloud concepts in the mobile network and hope to obtain an efficient network.

Efficient resource management [38] and orchestration [44], with and without mobility [45], to effectively manage the network slice to exploit is inherent flexibility will be paramount to servicing the vertical industries economically. The first challenge is for the graceful life cycle management of the virtual network functions (VNFs) and scaling of the system resources in a dynamic fashion. This is complicated by the fact that the VNFs are just pieces of software that can be instantiated anywhere in the network topology. The consequence of this is that the traditional protocol stack was designed without this flexibility of the location of the VNFs in mind. The interrelationships between the protocol stack and any instantiation has not been optimized in Release 15 which put certain constraints on the locational relationship between the VNFs. 3GPP is currently studying the issues and the optimization of the protocol stack for anywhere instantiation may be a feature in future releases. For example, there may need to be a design of a slice aware mobility management protocol to optimize the mobility challenges in network slicing [46].

The second challenge is to manage and orchestrate the resources such that the multiple network slices implemented on a common infrastructure can maintain isolation. The network would need to have resource elasticity to optimize the network sizing and resource consumption by exploiting statistical multiplexing gains [44] while maintaining isolation. Here, resource elasticity is defined as the ability of the wireless communication system to allocate and deallocate resources for each slice autonomously to scale the current available resources with the service demand gracefully.

The standards do not specify how the management and orchestration is done for network slicing. Such design is left up to the system designer. Part IV of this book will give some insights on how to achieve this in support of the vertical industry. Clearly, in order to customize the network for the service, joint resource allocation strategy that takes into account the significance of the resource to a particular service would be advantageous in supporting the diverse vertical applications on a common infrastructure [47]. Furthermore, to guarantee isolation, there is a desire to augment each virtual resource at instantiation with back up resources. This ensures that when failures occur, sufficient resources are available to maintain the service. In so doing, it becomes clear that there is a utilization tradeoff between reliability, isolation and efficient utilization of the resource. One method to circumvent this is to use pooled and shared backup resources [48]. The analysis showed that significant utilization improvement can be obtained.

One of the main challenges to slicing is an efficient method of exposing the capabilities of the network using network abstraction to the orchestration layer. This is compounded by the fact that the network is distributed and the service may span across several domains. There are many options for the system designer to implement the service orchestrator; ranging from simple distribution to federation. In [49] an orchestrator component was proposed that monitors and allocates virtual resources to the network slices and makes use of federation with other administration domains to take decisions on the end-to-end virtual service.

The distribution of the service over many domains has a profound impact on the security of the service. In the extreme case, the distributed infrastructure that the service runs on may come from different MNOs. The MVNO leases these resources in a dynamic fashion to fulfill its service requirement. 3GPP, in the development of the standards, states that non-management prerequisites "such as trust relationships between operators, legal and business related, to create such a slice are assumed to exist" [50] between the MNOs and MVNOs for multi-operator slice creation. Consequently, the open environment which enables the trading of resources, perhaps in the form of slices, is required to facilitate the isolation and scalability of the vertical services implemented over a shared infrastructure. In [51], using Factory of the Future as the exemplary use case, it was suggested that a 5G network slice broker in a blockchain can reduce the service creation time while autonomously handle network service request. The sharing of crosshaul capabilities in a 5G network via a multi-domain exchange was analyzed in [52] for the architectures proposed in the EU H2020 project. Another proposal, [53], is to use fair weighted affinity-based scheduling heuristic to solve the scheduling of micro services across multiple clouds.

Several reports demonstrating slice management in a 5G network have been reported in the literature. An overview of the different experimental SDN/NFV control and orchestration for the ADRENALINE test bed was presented [54]. The test bed can then be used for the development and testing of end-to-end 5G vertical services. In [55], the authors proposed an end-to-end holistic operational model following a top-down approach. They planned to realize the service operational framework within the MATILDA EU H2020 project.

1.6 Other Network Issues

The previous section discussed the significance of network slicing for the support of the vertical industry services. This section will examine a number of other issues that impact the vertical services such as VNF placement, MEC, and artificial intelligence (AI).

In a 5G network, where network function are virtualized and running on distributed hardware, the network topology to instantiate the VNF can have significant impact on the efficiency of the service. This problem is further compounded by the non-uniform nature of the service demand and the irregular nature of the network topology. This problem is not addressed in the standards because it is considered an implementation issue but is an extremely active area in the research literature. One solution [56] is to map the non-uniform distribution of signaling messages in the physical domain to a new uniform environment, a canonical domain, and then use the Schwartz–Christoffel conformal mappings to place the core functions. The analysis showed that the solution enhanced the end-to-end delay and reduced the total number of activated virtual machines. An affinity-based approach was shown in [57] to solve the function placement problem better than the greedy approach using the first-fit decreasing method. A multi-objective placement algorithm [58] only performs 5% less than the one obtained with the optimal solution for the majority of considered scenarios, with a speedup factor of up to 2000 times. The foregoing is not meant as an exhaustive review of the work on VNF placement but rather to show the vibrant research that can be used for supporting vertical services.

Another area of significant research activity is in distributed edge computing. The distribution of functions, possibly, over multiple cloud infrastructures, and the control and data traffic through these functions is known as service function chaining (SFC) [59]. An architecture for providing cost effective MEC and other services for the vertical industry was detailed and validated in [60]. An interesting problem, if there is now significant computation done at the edge, is that mobility would imply that there needs to be some mechanism to enable service migration at the edge. Network virtualization and distribution for data services over distributed enhanced packet core was discussed in [61]. Reference [62] developed the concept of a Follow Me Edge-Cloud leveraging the MEC architecture to sustain requirements of the 5G automotive systems. In principle the exact algorithm of linking the service to the cloud is up to the system designer, however, the standards provides the support to link the mobile service to the edge computing. Data plane distribution also has profound impact on the control plane side. To satisfy the low latency requirement, like the data plane, the control plane needs to be hierarchical in nature [63]. This then implies [64] that for an efficient mobile wireless system, certain data in the control plane need to be synchronized in a distributed architecture. This is particularly important because of the statefulness of the commercial wireless system.

One area of research that is getting a resurgence is using AI for network management. 3GPP has studied the usage of AI and is currently standardizing a network Data analytic function and its interface in the second phase of 5G (Release 16). The Internet Engineering Task Force (IETF) Autonomic Networking Integrated Model and Approach (ANIMA) working group is working on self-managing characteristics (configuration, protection, healing, and optimization) of distributed network elements, and adapting to unpredictable changes. The support of AI for vertical services is on its way. It should be noted that the standards will not standardize the AI methodology but rather will only standardize the interface to support AI functions. The interested reader can find details on the challenges and opportunities in [65].

1.7   Book Outline

It should now be clear that 5G is a network that was endowed, from the very beginning, with capabilities that allow for economical customization of the wireless network for different vertical industries; indeed, maybe to such a fine scale as each application. This chapter is the only chapter in Part I of this book. The rest of the book will elucidate what enables 5G to have such attributes and how it will revolutionize the communications within the vertical industries. It is organized as shown in the following.

Part II of this book is on deployments and business model opportunities and challenges. There is one chapter (Chapter 2) in this part. This chapter describes how the 5G network is designed to support a variety of vertical services from the operator point of view. The chapter introduces a variety of 5G services and their requirements. This chapter also gives a detailed presentation on the 5G network deployment architecture. Furthermore, this chapter also discusses a service-aware SON and some practical use cases. Performance benefits are also analyzed.

Part III of this book is on radio access technologies for 5G verticals. The discussions of the 5G standards here are necessarily laconic. The interested readers are encouraged to consult the more pedagogical discussion in [8]. There are three chapters (Chapters 3, 4, and 5) in this part.

Chapter 3 is on NR radio interface for 5G verticals. This chapter provides an overview of the 3GPP NR radio interface design and explains how it can be tailored to meet the requirements of different verticals. The chapter also covers advanced technologies for NR such as scalable OFDM numerology, flexible slot structure, massive multiple-input multiple-output (MIMO), beamforming, advanced channel coding, millimeter-wave deployment, spectrum aggregation, and dual connectivity and how these features enable various verticals.

Chapter 4 is dealing with one of the most important issues in millimeter-wave communications, i.e. the effects of dynamic blockage in multi-connectivity millimeter-wave radio access. This chapter provides a tutorial on modeling the dynamic blockage processes for 5G NR connectivity, and then shows how to improve the session connectivity in the presence of dynamic blockage.

Chapter 5 is on radio resource management techniques for 5G verticals. This chapter discusses radio access network resources, network slicing and its challenges for achieving efficient resource management; then it exposes the reader to the resource management approaches that can support and build efficient network slices for 5G verticals. Furthermore, resource allocation techniques and performance analysis are shown for virtual reality use case.

Part IV of this book is on network infrastructure technologies for 5G verticals. There are two chapters (Chapters 6 and 7) in this part.

Chapter 6 is on advanced NFV, SDN and mobile edge technologies for URLLC verticals. This chapter gives an overview of several URLLC vertical scenarios, their requirements and different deployment scenarios; it then discusses SDN, NFV and

5G core network functions to provide high precision networking to match the bandwidth, latency, and reliability targets of different URLLC applications.

Chapter 7 is on edge clouds complementing 5G networks for real time applications. This chapter describes the basics of edge cloud with respect to 5G networking and deployment, and SDN, NFV, and the need for disaggregation of control from data planes to provide the best practices at the edge where control from the cloud is collapsing with the central RAN control at the edge. The chapter also provides the state-of-the-art for edge cloud deployment options.

Part V of this book is on 5G key vertical applications. There are three chapters (Chapters 8, 9, and 10) in this part.

Chapter 8 is on 5G connected aerial vehicles. This chapter summarizes the needs and challenges to support aerial vehicles over current and future cellular networks, and provides a review of the research and development of drone communications in general from both academia and industry. This chapter also gives a discussion on the 5G challenges for aerial vehicle solutions and further work needed in this area.

Chapter 9 is on 5G connected automobiles. This chapter focuses on the value 5G brings to connected vehicles, specifically on high data rates, low latency, and edge computing features. This chapter concentrates on a couple of main use cases, vehicle platooning, and high definition maps, and how 5G and edge computing can assist to provide the best experience to the end-user.

Chapter 10 is on 5G for the industrial application. It will study the capabilities of the current release of 5G and its applicability to the smart factory use cases. The gaps to the requirement will be identified and how the industry is working to find solutions to close the gaps will be elucidated. The chapter concludes with a discussion on the spectrum situation of industrial use as well as some early trials and demonstrations.

References

1 Andrews, J.G., Buzzi, S., Choi, W. et al. (2014). What will 5G be? *IEEE Jounal on Selected Areas in Communication* 32 (6): 1065–1081.

2 Liu, J., Xiao, W., and Soong, A.C.K. (2016). Dense network of small cells. In: *Design and Deployment of Small Cell Networks* (eds. A. Anpalagan, M. Bennis, and R. Vannithamby), 96–120. Cambridge: Cambridge University Press.

3 Schwab, K. (2016). *The Fourth Industrial Revolution*. Geneva: World Economic Forum.

4 Word Economic Forum (2019). A New Era of Manufacturing in The Fourth Industrial Revolution: $7 Billion of Possibilities Uncovered in Michigan. https://www.weforum.org/whitepapers/a-new-era-of-manufacturing-in-the-fourth-industrial-revolution-7-billion-of-possibilities-uncovered-in-michigan.

5 General Motors (2019). U.N. Sustainable Development Goals. https://www.gmsustainability.com/unsdg.html (accessed 29 April 2019).

6 Ford (2018). Contributing to the UN SDGs. https://corporate.ford.com/microsites/sustainability-report-2017-18/strategy-governance/sdg.html (accessed 29 April 2019).

7 United Nations (2015). Transforming our World: The 2030 Agenda for Sustainable Development. https://sustainabledevelopment.un.org/post2015/transformingourworld/publication.

8 Lei, W., Soong, A.C.K., Jianghua, L. et al. (2020). *5G System Design: An End to End Perspective*. Cham: Springer International Publishing.

9 3GPP (2018). 3GPP Specification Series 38. http://www.3gpp.org/DynaReport/38-series.htm.

10 Global Data (n.d.). GSMA Intelligence. https://www.gsmaintelligence.com (accessed 20 August 2018).

11 Hung, M. (2017). *Leading the IoT*. Gartner Research.

12 Anderson, C. (2006). *The Long Tail*. Westport: Hyperion.

13 Campbell, K., Diffley, J., Flanagan, B. et al. (2017). The 5G economy: How 5G Technology will contribute to the global economy. https://cdn.ihs.com/www/pdf/IHS-Technology-5G-Economic-Impact-Study.pdf.

14 Word Economic Forum (2017). Digital Transformaton Initiative: Telecommunication Industry. http://reports.weforum.org/digital-transformation/wp-content/blogs.dir/94/mp/files/pages/files/dti-telecommunications-industry-white-paper.pdf.

15 Directorate-General for Communications Networks, Content and Technology (European Commission) (2017). Identification and quantification of key socio-economic data to support strategic planning for the introduction of 5G in Europe. https://publications.europa.eu/en/publication-detail/-/publication/ee832bba-ed02-11e6-ad7c-01aa75ed71a1.

16 Lee, N.T. (2019). Enabling opportunities: 5G, the internet of things, and communitiesofcolor.https://www.brookings.edu/research/enabling-opportunities-5g-the-internet-of-things-and-communities-of-color.

17 Tudzarov, A. and Gelev, S. (2018). 5G and software network paradigm. 23rd International Scientific-Professional Conference on Information Technology (IT).

18 Chiosi, M., Clarke, D., Willis, P. et al. (2012). Network Functions Virtualisation – An Introduction, Benefits, Enablers, Challenges and Call for Action. http://portal.etsi.org/NFV/NFV_White_Paper.pdf.

19 Ravindran, R., Chakraborti, A., Amin, S.O. et al. (2017). 5G-ICN: delivering ICN services over 5G using network slicing. *IEEE Communications Magazine* 55 (5): 101–107.

20 Rost, P., Mannweiler, C., Michalopoulos, D.S. et al. (2017). Network slicing to enable scalability and flexibility in 5G mobile networks. *IEEE Communications Magazine* 55 (5): 72–79.

21 Gavras, A., Denazis, S., Tranoris, C. et al. (2017). Requirements and design of 5G experimental environments for vertical industry innovation. 2017 Global Wireless Summit (GWS).

22 Elayoubi, S.E., Fallgren, M., Spapis, P. et al. (2016). 5G service requirements and operational use cases: Analysis and METIS II vision. 2016 European Conference on Networks and Communications (EuCNC).

23 Pateromichelakis, E., Bulakci, O., Peng, C. et al. (2018). LAA as a key enabler in slice-aware 5G RAN: challenges and opportunities. *IEEE Communications Standards Magazine* 2 (1): 29–35.

24 Mahmood, K., Mahmoodi, T., Trivisonno, R. et al. (2017). On the integration of verticals through 5G control plane. 2017 European Conference on Networks and Communications (EuCNC).

25 Pilz, J., Holfeld, B., Schmidt, A., and Septinus, K. (2018). Professional live audio production: a highly synchronized use case for 5G URLLC systems. *IEEE Network* 32 (2): 85–91.

26 3GPP (2018). TR 38.913 Study on scenarios and requirements for next generation access technologies v15.0.0. https://portal.3gpp.org/desktopmodules/ Specifications/SpecificationDetails.aspx?specificationId=2996.

27 NGMN (2015). NGMN 5G White Paper. https://www.ngmn.org/5g-white-paper/5g-white-paper.html.

28 Ji, H., Park, S., Yeo, J. et al. (2018). Ultra-reliable and low-latency communications in 5G downlink: physical layer aspects. *IEEE Wireless Communications* 25 (3): 124–130.

29 Simsek, M., Aijaz, A., Dohler, M. et al. (2016). 5G-enabled tactile internet. *IEEE Journal on Selected Areas in Communications* 34 (3): 460–473.

30 Alsenwi, M., Tran, N.H., Bennis, M. et al. (2019). eMBB-URLLC resource slicing: a risk-sensitive approach. *IEEE Communications Letters* 23 (4): 740–743.

31 Ordonez-Lucena, J., Ameigeiras, P., Lopez, D. et al. (2017). Network slicing for 5G with SDN/NFV: concepts, architectures and challenges. *IEEE Communications Magazine* 55 (5): 80–87.

32 Kalyoncu, F., Zeydan, E., and Yigit, I.O. (2018). A data analysis methodology for obtaining network slices towards 5G cellular networks. IEEE 87th Vehicular Technology Conference (VTC Spring.

33 Soenen, T., Banerjee, R., Tavernier, W. et al. (2017). Demystifying network slicing: From theory to practice. 2017 IFIP/IEEE Symposium on Integrated Network and Service Management (IM).

34 Schneider, P., Mannweiler, C., and Kerboeuf, S. (2018). Providing strong 5G mobile network slice isolation for highly sensitive third-party services. IEEE Wireless Communications and Networking Conference (WCNC).

35 Casetti, C., Chiasserini, C.F., Deiß, T. et al. (2018). Network slices for vertical industries. IEEE Wireless Communications and Networking Conference Workshops (WCNCW).

36 Rezende, P.H.A. and Madeira, E.R.M. (2018). An adaptive network slicing for LTE radio access networks. 2018 Wireless Days (WD).

37 Osseiran, A., Elloumi, O., Song, J., and Monserrat, J.F. (2018). Internet-of-Things (IoT). *IEEE Communications Standards Magazine* 2 (2): 70.

38 Richart, M., Baliosian, J., Serrat, J., and Gorricho, J.-L. (2016). Resouce slicing in virtual wireless networks: a survey. *IEEE Transactions on Network and Service Management* 13 (3): 462–476.

39 Foukas, X., Patounas, G., Elmokashfi, A., and Marina, M.K. (2017). Network slicing in 5G systems. *IEEE Communications Magazine* 55 (5): 94–100.

40 Kaloxylos, A. (2018). A survey and an analysis of network slicing in 5G networks. *IEEE Communications Standards Magazine* 2 (1): 60–65.

41 Sharma, S., Miller, R., and Francini, A. (2017). A cloud-native approach to 5G network slicing. *IEEE Communications Magazine* 55 (8): 120–127.

42 Ksentini, A. and Nikaein, N. (2017). Toward enforcing network slicing on RAN: flexibility and resource abstraction. *IEEE Communications Magazine* 55 (6): 102–108.

43 Sallent, O., Perez-Romero, J., Ferrus, R., and Agusti, R. (2017). On radio access network slicing from a radio resource management perspective. *IEEE Wireless Communications* 24 (5): 166–174.

44 Gutierrez-Estevez, D.M., Gramaglia, M., De Domenico, A. et al. (2018). The path towards resource elasticity for 5G network architecture. IEEE Wireless Communications and Networking Conference Workshops (WCNCW).

45 Zhang, H., Liu, N., Chu, X. et al. (2017). Network slicing based 5G and future Mobile networks: mobility, rcsourcc management, and challenges. *IEEE Communications Magazine* 55 (8): 138–145.

46 Li, X., Samaka, M., Chan, H.A. et al. (2017). Network slicing for 5G: challenges and opportunities. *IEEE Internet Computing* 21 (5): 20–27.

47 Narmanlioglu, O., Zeydan, E., and Arslan, S.S. (2018). Service-aware multi-resource allocation in software-defined next generation cellular networks. *IEEE Access* 6: 20348–20363.

48 Yeow, W.-L., Westphal, C., and Kozat, U.C. (2011). Designing and embedding reliable virtual infrastructures. *ACM SIGCOMM Computer Communication Review* 41 (2): 57–64.

49 Li, X., Mangues-Bafalluy, J., Pascual, I. et al. (2018). Service orchestration and federation for verticals. IEEE Wireless Communications and Networking Conference Workshops (WCNCW).

50 3GPP (2018). TR 28.801 v15.1.0: Study on management and orchestration of network slicing for next geration networks. http://www.3gpp.org/ftp//Specs/archive/28_series/28.801.

51 Backman, J., Yrjölä, S., Valtanen, K., and Mämmelä, O. (2018). Blockchain network slice broker in 5G: Slice leasing in factory of the future use case. *Internet of Things Business Models, Users, and Networks*, Copenhagen, Denmark (23–24 November 2017), IEEE.

52 Contreras, L.M., Bernardos, C.J., Oliva, A.D.L., and Costa-Perez, X. (2017). Sharing of crosshaul networks via a multi-domain exchange environment for 5G services. IEEE Conference on Network Softwarization (NetSoft).

53 Bhamare, D., Samaka, M., Erbad, A. et al. (2017). Multi-objective scheduling of micro-services for optimal service function chains. IEEE International Conference on Communications (ICC).

54 Muñoz, R., Vilalta, R., Casellas, R. et al. (2017). Integrating optical transport network testbeds and cloud platforms to enable end-to-end 5G and IoT services. 19th International Conference on Transparent Optical Networks (ICTON).

55 Gouvas, P., Zafeiropoulos, A., Vassilakis, C. et al. (2017).Design, development and orchestration of 5G-ready applications over sliced programmable infrastructure. 29th International Teletraffic Congress (ITC 29).

56 Laghrissi, A., Taleb, T., and Bagaa, M. (2018). Conformal mapping for optimal network slice planning based on canonical domains. *IEEE Journal on Selected Areas in Communications* 36 (3): 519–528.

57 Bhamare, D., Samaka, M., Erbad, A. et al. (2017). Optimal virtual network function placement in multi-cloud service function chaining architecture. *Computer Communications* 102: 1–16.

58 Arouk, O., Nikaein, N., and Turletti, T. (2017). Multi-objective placement of virtual network function chains in 5G. IEEE 6th International Conference on Cloud Networking (CloudNet).

59 Halpern, J. and Pignataro, C. (eds.) (2015). RFC 7665: Service Function Chaining (SFC) Architecture. https://www.rfc-editor.org/info/rfc7665.

60 Vilalta, R., Lopez, V., Giorgetti, A. et al. (2017). TelcoFog: a unified flexible fog and cloud computing architecture for 5G networks. *IEEE Communications Magazine* 55 (8): 36–43.

61 Kaippallimalil, J. and Chan, H.A. (2014). Network virtualization and direct Ethernet transport for packet data network connections in 5G wireless. IEEE Global Communications Conference.

62 Aissioui, A., Ksentini, A., Gueroui, A.M., and Taleb, T. (2018). On enabling 5G automotive systems using follow me edge-cloud concept. *IEEE Transactions on Vehicular Technology* 67 (6): 5302–5316.

63 Sama, M.R., Contreras, L.M., Kaippallimalil, J. et al. (2015). Software-defined control of the virtualized mobile packet core. *IEEE Communications Magazine* 53 (2): 107–115.

64 Kaippallimalil, J., Mademann, F., Shiyong, T. et al. (2015).Data distribution and synchronization in next generation mobile core network. IEEE Conference on Standards for Communications and Networking (CSCN).

65 Mwanje, S., Decarreau, G., Mannweiler, C. et al. (2016).Network management automation in 5G: Challenges and opportunities. IEEE 27th Annual International Symposium on Personal, Indoor, and Mobile Radio Communications (PIMRC).

Part II

5G Verticals – Deployments and Business Model Opportunities and Challenges

2

5G Network for a Variety of Vertical Services

Jin Yang

Verizon Communications Inc., Walnut Creek, CA, USA

Abstract

This chapter illustrates 5G services, deployment architecture, service-aware autonomous network operation and management. Thus, it describes how a commercial wide-area network can support vertical markets, and how a vertical market can utilize the means in a commercial network to address their needs. The chapter provides a summary of network requirement for typical use cases. The 5G flat network architecture and standardized high-layer radio access network (RAN) split F1 interface allow more flexible service delivery and broader ecosystem. The chapter proposes a multi-access edge computing network architecture with virtualized RANs. An autonomous intelligent service-aware self-organizing network (SON) is required for the 5G network with control functions at radio, baseband distributed unit, higher layer central unit, and various edge and core servers. Operator and vertical provider play an essential and more active role to ensure smooth and coordinated operations of various SON features among a large number of configurable components.

Keywords 5G flat network architecture; edge computing network architecture; radio access network; self-organizing network; vertical markets

The fifth generation (5G) of cellular mobile technology is transforming human society. The gigabit-per-second ultra-wide broadband service is happening now [1]. 5G is the enabler for a fully connected world with communications of everything. It will support a variety of services from various vertical industries.

5G wireless networks have fundamentally evolved from a connectivity-based network to an intelligent service delivery platform. 3GPP has successfully

---

5G Verticals: Customizing Applications, Technologies and Deployment Techniques, First Edition.
Edited by Rath Vannithamby and Anthony C.K. Soong.

completed the first implementable 5G New Radio (5G-NR) specifications in Release 15 [2]. This standard has enabled full-scale deployment of 5G-NR. The standard body is continuing to enhance end-to-end network functionalities in 5G core (5GC) [3]. Those are crucial for operators to further explore the advanced capabilities for consumers, enterprises, and different vertical market segments. The 5GC supports end-to-end network slicing and differentiated quality of service (QoS) awareness from radio, transport to core and application server to serve various vertical markets. The vertical markets should also leverage 5G-NR and 5GC new functionalities for more efficient and reliable services.

Intensive researches have been conducted on this new architecture for various applications. The European Telecommunications Standards Institute (ETSI) issued a white paper on multi-access edge computing (MEC) in 5G [4]. The 5G architecture enables a more flexible software-based network on a distributed cloud platform. The tradeoffs among latency and reliability with spectral efficiency and coverage are analyzed for tactile internet services [5]. Latency critical Internet of Things (IoT) applications and requirements are studied [6]. A more advanced network management system (NMS) [7] with intelligent Self-Organizing Network (SON) is required for this network [8]. Machine learning (ML) is necessary to control this sophistical system with a large amount of adjustable parameters and service performance metrics [9, 10]. Some practical applications of the ML are presented by researchers and operators [11, 12]. SON and Cloud-RAN are shown to be essential to enable ultra-reliable and low-latency communications (URLLC) on top of enhanced mobile broadband (eMBB) and massive machine type communication (mMTC) [13, 14].

This chapter illustrates 5G services, deployment architecture, service-aware autonomous network operation and management. Thus, it describes how a commercial wide-area network can support vertical markets, and how a vertical market can utilize the means in a commercial network to address their needs. A summary of network requirement for typical use cases is provided. The 5G flat network architecture and standardized high-layer Radio Access Network (RAN) split F1 interface allow more flexible service delivery and a broader ecosystem. A MEC network architecture with virtualized RANs is proposed. An innovative hybrid 3-tier SON is exploited to incorporate ML. Specifically, a middle-tier SON (mSON) is defined to take advantage of the F1 interface, in addition to conventional centralized SON (cSON) and distributed SON (dSON) functionalities. This is essential to support less than millisecond ultra-low latency applications at 99.999% reliability. Mobile wireless network is transforming into an intelligent multiple-purpose autonomous 5G network for a variety of services.

The rest of the chapter is organized as follows: a variety of 5G services and their requirements will be introduced in Section 2.1; the 5G deployment architecture is presented in Section 2.2; a service-aware SON is discussed in Section 2.3; and some practical use cases and performance benefits are provided in Section 2.4. A summary is given in Section 2.4.

2.1 5G Services

A 5G network can support a variety of 5G services to enable a fully connected world with communications of everything. A traditional cellular network is specifically built for voice and Internet data services among human beings. The generations of access evolution from 2G to 4G has emphasized spectral efficiency. The 5G wireless network is fundamentally evolving from human oriented tele-communications to a connection of everything. In particular, three fundamental categories of services are defined to guide the development of the standards from the very beginning: eMBB, URLLC, and mMTC [15].

The three types of basic services and their combinations are changing lives and society. Table 2.1 illustrates examples of 5G services and their requirements.

The use cases in eMBB include personalized mobile broadband access for smartphone, fixed wireless communications as an alternative to cable communications, and high-fidelity video streaming. It requires the average throughput increase of 100 times from 100 Mbps up to 10 Gbps. The reliability is also improved significantly from the traditional 90 percentile to 97 percentile on mobile broadband, the same reliability as wireline at 99.999% for fixed wireless.

The URLLC consist of industrial IoT with remote controlled factory robotic operations for medical surgery or industry machines. It can also support remote controlled autonomous driving, and emergency real-time communications. It will further demand lower latency than 50 ms with reliability at 99.999–99.9999%. mMTC will support wide-area low power services, such as sensors in wineries and oil wells, parking, water and electric meters, and tracking devices across vertical markets. It requires millions of connections per cell with more than 10 years of battery life.

The Slice/Service Type (SST) in Table 2.1 is specified in TS23.501 [3] with value 1, 2, and 3 corresponding to eMBB, URLLC, and MTC, respectively. The requirement and enabler of each service type will be further discussed below.

2.1.1 Enhanced Mobile Broadband

eMBB is driven by exponential growth of traffic, as well as demands for immersive augmented reality (AR)/virtual reality (VR) and ultra-broadband wireless communications. Those services require average throughput in hundreds of Mbps, with peak throughput above gigabit per second.

5G-NR supports much broader frequency bandwidth in hundreds of MHz, with a large number of carrier aggregations. Massive MIMO (multiple-input multiple-output) with more than eight layers of data streams over multiple users simultaneously enables high spectral efficiency and peak data rate. Flexible frame structure with various subcarrier spaces allows dynamic spectrum sharing from hundreds of MHz to beyond 50 GHz with paired and unpaired spectrum.

Table 2.1 Example of 5G use cases and requirements.

Use cases	SST	Description	Traffic pattern	Availability	Average throughput	Latency/ Jitter	#Connections per cell	Other requirements
eMBB	1	Enhanced Mobile Broadband	Variable rate	0.97	0.1–1 Gbps	<200 ms	500+	UE speed up to 500 kmph
eMBB	1	Wireless backhaul (fixed)	Variable rate	0.999 99	0.5–10 Gbps	<50 ms	<100	Fixed access
eMBB	1	Secure channel for public safety	Variable rate	0.9999	100–250 Mbps	<150 ms	100–500	Authentication, security
eMBB	1	Multi-media operation (AR/VR)	Constant rate	0.9999	100–250 Mbps	<50 ms	<100	Stringent jitter and latency
URLLC	2	Industrial IoT	Variable rate	0.999 99	Up to 250 kbps	<10 ms	500	High reliability
URLLC	2	Vital sign	Variable rate	0.999 99	Up to 100 kbps	<20 ms	500	High reliability
URLLC	2	Autonomous driving	Variable rate	0.999 99	Up to 10 Mbps	<10 ms	100	UE speed up to 300 kmph
mMTC	3	Sensors	Variable rate	0.999	Up to 1 Mbps	>1 h	300000	No mobility, 10+ year battery
mMTC	3	Utility meters	Variable rate	0.999	Up to 100 kbps	>1 wk	300000	No mobility, 10+ year battery
mMTC	3	Tracker	Variable rate	0.999	Up to 100 kbps	>10 s	300000	10+ yr battery, extended coverage

The low-band in sub-6 GHz can be utilized for wide area services, with high-band for ultra-dense small cells or fixed wireless applications. 5G wireless needs to support licensed spectrum, and leverage unlicensed and shared spectrum to increase throughput and complement user performance.

Millimeter wave (mmWave) with more than 200 MHz bandwidth is critical for average user throughput larger than 1 Gbps in a loaded dense network with cell radius less than 100 m. The mmWave relies on antenna beam management to mitigate sensitivity to rain and fog. A massive antenna system can form and track user specific narrow-beam to ensure service quality and reliability. This can mitigate a relatively large propagation loss and recover from blockages.

Idle and connected mode mobility management is essential to ensure consistent user experience across the network. These are challenging in a heterogeneous network with large range of inter-site distances and frequency bands. NR supports mobility without reconfiguration, and make-before-break handover with multiple connections. It will reduce the handover interruption time to zero from current Long-Term Evolution's (LTE's) minimal 25 ms.

5G will bring in more flexibility on frequency and time allocation, and allow flexible uplink and downlink multiplex to improve spectrum utilization. The 5G NR in Release 15 will support more flexible Time Division Duplex (TDD) resource allocations, so each user can have individually specified downlink and uplink time slots. Full dynamic Frequency Division Duplex (FDD) and TDD allocation, as well as full duplex radios are expecting to improve Integrated Access Backhaul further.

In summary, ultra-wide bandwidth and network densification, with flexibility resource sharing among TDD/FDD, licensed/unlicensed/shared spectrums will fulfill the requirement for enhanced broadband. The mobility will be further enhanced with multiple transmission points, idle and connected mobility management with zero interruptions.

2.1.2 Ultra Reliable and Low Latency Communications

URLLCs expand the mobile cellular network to an entire new category of services. Flexible frame structure reduces the latency, while the massive amount of transmission points increases the reliability.

The 5G NR subcarrier spacing (SCS), Transmission Time Interval (TTI) and cyclic prefix are flexible and configurable based on deployed spectrum and service requirement [2]. It supports a shorter mini-slot at µs level, and larger SCS for a delay sensitive mission critical application with a finer resource block granularity. It also allows a narrow bandwidth and smaller SCS for massive IoT connections to extend coverage, reduce device complexity and power consumption.

The NR frame structure supports a self-contained transmission, hence it reduces overall transmission latency. The reference signals required for data demodulation are included in the given slot or beam. It also supports well confined transmissions in time and frequency, avoids the mapping of control channels across full system bandwidth. It avoids static timing relation across slots and different transmission directions, removes the resource allocation limitation of predefined transmission time. 5G-NR can achieve less than 1 ms latency.

NR supports $\pi/4$-BPSK modulation for DFT-s-OFDM to extend coverage and reliability. It supports low-density parity-check (LDPC) codes for the data channel and Polar codes for the control channel. The LDPC has shown much superior performance at a wide range of coding rates. It supports high peak rate and low latency at high coding rate, high coding gain with high reliability at low coding rate.

The massive amount of transmitters and receivers, either collocated by massive MIMO or non-collocated from multiple radio points, enables a reliable network to take full advantage of spatial diversity [16]. Tight time and frequency coordination among those multiple transmission points also further increases network reliability and resilience.

Thus, the new frame structure, modulation and coding schemes, as well as spatial, time, and frequency diversities enable the URLLC. Industrial IoT can utilize URLLC features to support high reliable IoT use cases. This can be robotics in factory assembly lines, or remote surgery at a patient's home. Vehicle-to-everything (V2X) type of services will require URLLC as well for remote controlled vehicles.

2.1.3   Massive Machine Type Communications

Connections beyond human beings have driven the growth of the wireless network as the penetration rate for humans has saturated. Verizon's State of the IoT Market – 2017 report [17] has predicted a double digital growth and expected game-changing 5G services. Traffics and revenue for IoT technologies have been more than doubled in the past year. Intensive field tests [18] have also demonstrated distinguished radio characteristics of those IoT applications.

5G-NR is addressing mMTC from three aspects: integrated LTE-M (LTE category M1) and NB-IoT (narrowband IoT) from legacy LTE, Industrial IoT and NR-IoT leveraging 5G-NR interface.

5G-NR SCS and frame structure allow orthogonality of NR with LTE-M and NB-IoT, which is original designed to be compatible with LTE. As a result, LTE-M and NB-IoT could be supported inside the same 5G-NR frequency, independent of LTE, as shown in Figure 2.1. This is essential as LTE-M and NB-IoT have just taken off in double-digit growth in the commercial network with a long customer contract commitment. LTE-M and NB-IoT can be deployed standalone or in-band. They will sustain much longer life than LTE.

Industrial IoT would require reliable communications supported by URLLC. NR-IoT enables an expanded new category of IoT services. It can take advantage

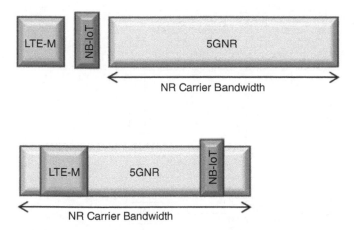

Figure 2.1 LTE-M and NB-IoT coexisting with 5G-NR.

of the flexible and dynamical frame structure of NR to allow wide-bandwidth, smaller SCS, flexible combination on time slots and frequency resource blocks. It will more efficiently schedule time and frequency radio resource based on better channel feedback information. NR-IoT could support higher data rate than LTE-M while keeping low device cost with extended coverage and battery life.

Extended LTE-M, NB-IoT, and NR-IoT provide a rich set of options for massive Machine type of services in the 5G family. Together with eMBB and URLLC, 5G services can cover a wide range of applications. Verticals can classify their use cases into a combination of those three fundamental 5G service categories to speed up execution.

2.2 Networks

5G network architecture [3] is evolving from a point-to-point architecture to a service-oriented architecture. This enables a more refined standardized virtualization platform, allowing intelligence to allocate features and services at core and various edges to support diversified applications.

In 5GC, user equipment (UE) connection to RAN is controlled by Access and Mobility Management Function (AMF) through NG-C interface as illustrated in Figure 2.2. The AMF includes the network slice selection functionality. User traffic data is connected to Data Network (DN) through User Plane Function (UPF) through NG-U interface, which is controlled by Session Management Function (SMF). Comparing with the Mobility Management Entity (MME) in Evolved LTE Packet Core (EPC), the separation of access control plane (CP) and session management of user plane (UP) in 5GC provides greater flexibility and scalability, particularly for CP only traffic or non-IP session applications. Serving and Packet Gateway

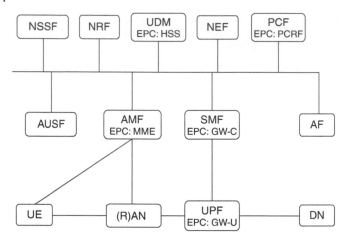

Figure 2.2 3GPP 5G architecture.

functions in EPC are separated as GW-C and GW-U, supported by SMF and UPF, respectively. Thus, it allows independent scalability, evolution and flexible deployments, e.g. centralized location or distributed (remote) location.

The Policy Control Function (PCF) provides a policy framework incorporating network slicing, roaming and mobility management, similar to the Policy and Charging Rules Function (PCRF) in EPC in LTE. Application Function (AF) requests dynamic policies and/or charge control. The Unified Data Management (UDM) stores subscriber data and profiles, similar to Home Subscriber Server (HSS) in LTE.

2.2.1   5G Network Architecture

The 5G network can be deployed as a standalone (SA) network based on 5G core or leverage existing EPC as a non-standalone (NSA), as illustrated in Figure 2.3. 5G-NR will be deployed with NSA first to leverage existing core network and extensive LTE coverage. Thus the initial services will be eMBB, followed by 5GC for additional services and capabilities.

The RAN and the core network components, network management and controller, as well as application servers are based on more general-purpose hardware, dynamically configurable under software-defined networking (SDN), as illustrated in Figure 2.4. The choice of general-purpose hardware and thereby the cloud implementation facilitates operators' provisioning of various services under a common physical platform, enabling resource sharing, service integration, and customization.

Differentiated services can be supported through end-to-end network slicing defined in 3GPP [3]. The network slicing allows dedicated virtual networks with

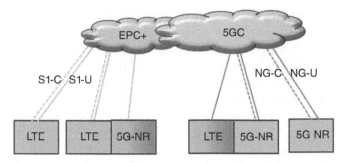

Figure 2.3   NSA and SA architecture.

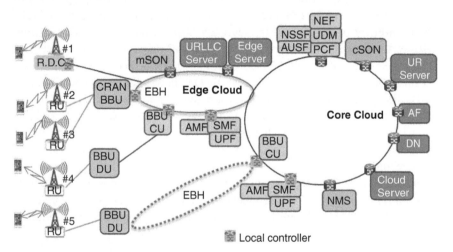

Figure 2.4   5G commercial network deployment architecture.

functionality specific to the service or customer over a common network infrastructure. Each network can be identified by Network Slice Selection Assistant Information (NSSAI). Network Slice Selection Function (NSSF) assists in the selection of suitable network slice instances (NSIs) for users and services. These network functions (NFs) and services are registered in a Network Resource Function (NRF). Network Exposure Function (NEF) acts as a centralized point for service exposure and authorizing all access requests from external systems. AMF is responsible for access authentication and authorization, mobility management and termination of CP interface and Non-Access Stratum (NAS). SMF supports session management, IP address allocation and management, selection and control of UPF, and QoS management.

A network slice corresponds to a dedicated set of resources that can meet a specific performance. A device can be simultaneously connected to multiple network slices. The NSSAI can be provided by UE or directed by a network controller. The NSSAI comprises SST and Slice Differentiator (SD). The SST refers to the expected network behavior in terms of features and services, as illustrated in Table 2.1. The SD is optional information that allows further differentiation by service providers.

Within each network slice, a packet data session could be further differentiated by a QoS flow. The QoS flow is identified by a QoS Flow ID (QFI). Each QoS flow is associated with 5G QoS Indicator (5QI), Allocation and Retention Priority (ARP), and Transport QoS marks (e.g. diffserv). On downlink, UPF uses policy from PCF and SMF to identify flows and adds QFI tag that RAN maps to data radio bearers (DRBs). On uplink, UE uses either signaling or "reflective" learning approach to learn and map QFI to DRBs. Hence, RAN and UPF police DRB mapping and QFI usage accordingly. In the example in Figure 2.5, UE1 has only one network slicing, severed by the network slice 1 through RAN and 5GC Core, while UE2 has slices 1 and 2, served by the RAN and 5GC core slices 2 and 3, respectively. Within each RAN slice, the radio resource is controlled by QoS flow, with a slice-aware scheduler. The mapping of slicing and QoS flow is essential for service quality.

The variety of services can be supported by distinguished network slicing, either UP or CP functions at far edge, edge center, or cloud data center. Network slicing implementation is end-to-end with different instances across radio, transport, core, edge and central clouds.

SDN and network function virtualization (NFV) virtualize the core network elements and functions in each slice to meet its own requirement. Core network components on both CP and UP can be pushed toward radio access point, edge cloud or

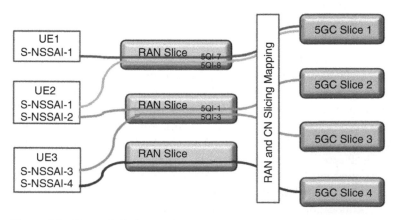

Figure 2.5 Network slicing.

core cloud with various levels of latencies. CP data can be directly carried over AMF. Packets over UP can be transmitted over UPF controlled by SMF and AMF.

In the RAN, slicing can be built on physical radio resources (e.g. transmission point, spectrum, time) or on logical resources abstracted from physical radio resources, autonomously configured and optimized using SON. The radio access functions can also be physically located at various edges for maximal radio efficiency. This includes traditional integrated base station, or separated radio and baseband units. The standardized interfaces allow more splitting combinations among radio unit (RU), distributed unit (DU) and central unit (CU), as illustrated in Figure 2.6. F1 interface between DU and CU, as well as E1 interface between CU control plane (CU-CP) and CU user plane (CU-UP) are specified in 3GPP. A common CU-CP will control various DUs in one next generation NodeB (gNB), select an AMF based on required NSSAIs, and direct UP traffic to various CU-UPs at different locations depending on latency and service requirements.

The commercial deployment architecture illustrated in Figure 2.4 shows several deployment options for cell sites based on service requirement. The cell site #1 has integrated radio, digital, and CUs directly connected to edge cloud, while sites #2 and #3 have RUs only connected to Centralized RAN (CRAN). The cell sites #4 and #5 have radio and digital units connected through F1 interface to edge and core clouds, respectively.

In the case of URLLC, the UP/CP functions and application servers could be located at edge cloud or even at access nodes to meet the latency requirement of even less than 10 ms, without dependency on the rest of the operator's transports and core networks. AMF/SMF/UPF and URLLC servers can be implemented physically or virtually at customer sites or virtually through operator data centers. Ultra-Reliable server that does not require stringent latency could reside at core cloud.

Figure 2.6 3GPP 5G-NR RAN architecture and interfaces.

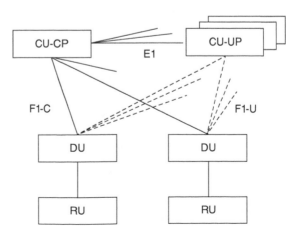

While the network is transforming to a service-oriented architecture, NMS is also evolving from traditional pure network performance monitoring to a fully autonomous 5G network. Various self-configurations, self-optimizing, and self-healing mechanisms enable zero-touch network provisioning and minimize human intervention from installations to commercial operations. Those SON mechanisms will be discussed further in the following sections.

5G network architecture allows one physical network to be sliced logically into multiple virtual isolated networks. Each slice can be optimized independently to serve a specific vertical application with various QoS flows. Thus, 5G networks provide a high degree of flexibility to enable MEC and virtualized radio access for a variety of services.

2.2.2 Multi-Access Edge Computing Network

The virtualized network and application servers can be tailored to a variety of vertical markets. It allows operators to create new value-added services for their customers with less dependency on the underlying hardware, making it transparent across the entire network. It enables low-latency and ultra-reliable services by pushing core network components and servers toward edge for MEC.

Cloud servers enhance network efficiency through resource pooling, and improve time-to-market for innovative services. However, they also introduce latency and uncertainty on routing. Edge servers enhance the latency but increase the cost and complexity by adding computing, caching and other service functions at the edge. They also introduce challenges for mobility. Thus, it is essential to have a local controller at various tiers of edge clouds to determine the proper routing and content transitions.

An agile way to direct traffics based on a loss function of latency and serving-node cost is proposed. The latency driven routing and service platform are enabled through a multi-tier MEC which includes a far-edge server, edge-cloud server, and a wide-area core cloud server. The controller for this service platform can be co-located with integrated base station at far-edge, CRAN at edge cloud, or wide-area core cloud as illustrated in Figure 2.4. Device service profile and routing authorization are used to direct a service.

The loss function is derived from latency, and service node cost for a device at its specific location i, denoted as $Delay_i$ and $Cost_i$, respectively. The cost includes the network resource needed for:

1) Content preparation, e.g. content availability, content popularity, and replication of content across localized MEC servers.
2) Transport cost associated with UE location, mobility.
3) Caching and computing.

Thus, the selected route j ($S_{\text{route}j}$) is the lowest loss function route, that is

$$S_{-\text{route}j} = \min\{\text{Cost}_i, \; i = 1, \; ..., N\}, \; while \; \text{Delay}_i < \text{Delay Target}$$

The selected route shall meet the targeted latency with minimal loss functions, as illustrated in Figure 2.7. Various hider-layers can be utilized. Supervisory learning can be introduced first before autonomous operation. The co-location of 5GC with access node can achieve latency of less than milliseconds and reliability at 99.999%. The default cloud based 5GC enables more efficient resource pooling with eMBB in latency beyond 50 ms.

Mobility for MEC is challenging as the server is localized. The 5GC has enhanced session and service continuity (SSC) through various means; the three SSC modes are [3]:

- SSC mode 1: the network preserved the continuity service provided to the UE with the same PDU IP anchor maintained for the duration of UE attachment. This is similar to 4G LTE.

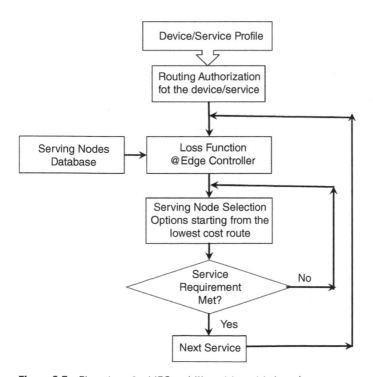

Figure 2.7 Flowchart for MEC mobility with multi-tier edges.

- SSC mode 2: the network may release the connectivity service delivery to the UE. The IP address is not preserved, and the existing session will be ended.
- SSC mode 3: A new PDU IP anchor is established when UE moves in the network. The previous PDU IP anchor is maintained until a new PDU IP anchor point is established. The new connection will be established before the existing session ends.

For use plane application, 5GC can also support breakout UPF using Uplink Classifier Rules for traffic steering to route selective traffic to the local MEC platform for low-latency applications directly. CP applications need to be directed by NSSF through NSSAI.

5GC enables MEC at various radio access and core nodes for a variety of applications. It also supports various traffic steering, session and service continuality.

2.2.3 Virtualized Radio Accesses

RAN virtualization can maximize hardware and radio resource efficiency through pooling and tightly coordinated radio resource allocation. Virtualizations broaden the 5G ecosystem with more innovated solutions from more players. Radio access virtualization is enabled through standardized interfaces between CU and DU. Different RAN functional CU-DU splits are under intensive investigations by both academics and industry fora. The CU can also be separated into CP and UP to allow more dynamic resource control.

The splitting of RAN architecture into CUs and DUs needs to cater for different deployment scenarios and challenges, including supporting the introduction and evolutions of various RAN capabilities, such as coordinated/joint processing, massive MIMO, advances receivers, etc. It is also required to facilitate new application developments and allow business innovations, such as eMBB, cellular V2X services, mMTC devices, and so on. The splits should further attempt to minimize front-haul throughput requirements scaling as a function of user throughputs.

3GPP has defined various split options between CU and DU based on the reference models in [19, 20]. The most widely studied split options are illustrated in Figure 2.8.

Option 2 has Radio Resource Control (RRC) and Packet Data Convergence Protocol (PDCP) in CU; while Radio Link Control (RLC), Medium Access Control (MAC), Physical layer and RF are in DU. This has been standardized as F1 interface in 3GPP [21]. Option 7 splits within Physical layer, while Option 8 is the traditional Physical layer and RF layer split as typically defined by the Common Public Radio Interface (CPRI) interface. Various low-layer splitting has been studied in 3GPP [22].

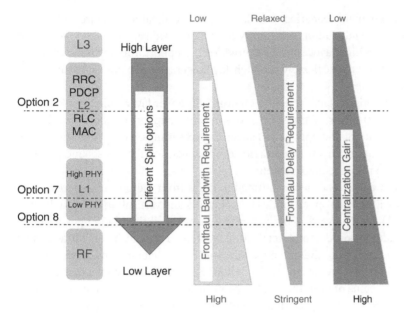

Figure 2.8 RAN splitting and performance tradeoff.

The standardized splitting can broaden the ecosystem as illustrated in Figure 2.4. Traditional macro-cell product from one single supplier can be broken into various radio, distributed, and CU products with various combinations. Those function elements can also be built on general hardware to allow virtualized radio access at RU, edge cloud, or center cloud.

The different split options determine the tradeoff between front-haul bandwidth and latency requirements as well as the centralization gains due to improved cell coordination capabilities as indicated in Figure 2.8. The lower the split point the higher the front-haul bandwidth required, the more stringent the font-haul delay is needed, and hence, the higher the centralized spectral efficiency gain. These split options and their variations present different deployment opportunities and challenges. While specifying more split options would mean greater deployment flexibility, inter-operability and building of the ecosystem becomes more difficult. Therefore, it is generally agreed that the standardization of two to three split options should be a good compromise between deployment flexibilities and development complexities.

High-layer split is particularly attractive for mmWave due to its moderate transport bandwidth requirements. Specifically, F1 interface has similar bandwidth requirements as user data rates and can tolerate a more relaxed latency of 1.5–10 ms compared with other lower layer split options. Hence, it supports a wide range of deployment scenarios. Low-layer split is critical for resource allocation and interference mitigation with tightly coordinated transmission and receive for sub-6 GHz. The high-layer split also provides the opportunity for network management to direct traffic flexibly at a high-layer aggregation point based on service requirements.

Option 2 for higher layer split and Option 7 for lower layer split provide logical compromises in terms of transport requirements, support of coordinated RUs, RAN virtualization, and application specific QoS settings. This encourages more innovative solutions in advanced radio components, sophistic baseband processing, and radio resource management.

Virtualized end-to-end network functionalities from radio to transport to core and server will provide opportunities for operators and verticals to define a distinct and isolated logical network of individual use cases in real-time, built upon a shared physical wide-deployed commercial infrastructure. This can be an ultra-reliable low latency service at the far edge for superior reliability and performance. Alternatively, it can leverage the wide-area 5G NR eMBB service for a cost efficient way of connections of everything. The 5G network can be readily configured as a private low-cost and high-performance network for verticals. Thus it provides enormous business opportunities for operators and verticals.

2.3 Service-Aware SON

An autonomous intelligent service-aware SON is required for the 5G network with control functions at radio, baseband DU, higher layer CU, and various edge and core servers. The flexible 5G architecture and radio access schemes increase the complexity of network management, with increased spectrums and network densifications. Standardized CU-DU splitting also provided more opportunities to control radio transmissions. The large intensified scope of network coordination, configuration and management has evolved to a level beyond which engineers can handle.

The performance data of the 5G NF are fundamental for network monitoring, assessment, analysis, optimization, and assurance. A variety of communication service instances are provided by multiple NSIs. The different parts of an NSI are grouped as Network Slice Subnets (e.g. RAN, 5GC, and Transport) allowing the lifecycle of a Network Slice Subnet Instance (NSSI) to be managed independently from the lifecycle of an NSI.

A SON framework is illustrated in Figure 2.9. SON consists of self-configuration, self-optimization, and self-healing. It is driven by consumers and services. The 5G SON function consists of the provisioning management services [23], performance data file services [24], and fault supervision data report management

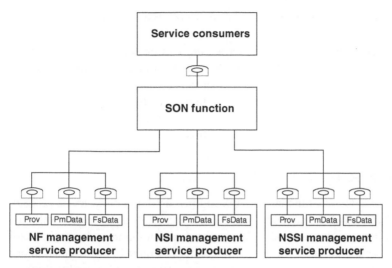

Prov: Provisioning management service
PmData: Performance data reporting management service
FsData: Fault supervision data reporting management service
Service consumers: For example. IoT applications, edge computing applications, ... etc

Figure 2.9 Service consumer-based SON.

service [23] to collect the management and network data from the management services produced for NF, NSIs, and NSSI. Cell trace data are also very critical to improve SON performance [25]. Based on the collected data, the SON function will analyze the network behavior, status, and traffic pattern, among others, to determine the actions needed to optimize the networks.

An intelligent service-aware SON is required to optimize the services delivery and network performance, such as reliability and latency, autonomously in real-time. Various ML algorithms are essential to take advantage of a large amount of big data, engineering rules and learning. The initial parameters and network configuration should be settable by experts to enable ML achieving local optimal points faster, and better approaching global optimum points. The ML inputs, hidden layers and loss functions are highly dependent on various services and network key performance indicators (KPIs). ML algorithms shield the complexity of physical and virtual NFs from higher layers to aid engineers.

Various ML algorithms in terms of supervised learning, unsupervised learning, and reinforcement learning are described in ML surveys [9, 10]. The dominated algorithm is Q-learning using the reinforcement approach, while the algorithm is driven by real-time network performance data reinforced by feedback from past execution results. Although limited applications have shown some preliminary successful use cases, current SON is still behaving less intelligent than experienced engineers. More intelligent and further inputs from end-to-end services and consumer device performance metrics need to be included in those algorithms. Combining machine-level insights with human insights should be exploited further to balance human and machine decision making.

Service-aware SON is needed beyond traditional distributed function at eNB and centralized function at NMS level to form an intelligent hybrid SON with radio control functions at core and various edges.

2.3.1 5G-NR SON Control

5G-NR can enable SON functionality beyond just eNB and NMS levels [26–30]. Traditional SON with functions at base station and NMS has proven a valuable component in network operations. More sophistical autonomous network control is required for the 5G network to ensure overall optimized performance under operator's policy control. A finer resolution under NMS level is possible through the high-layer CU-DU splitting F1 interfaces defined in 3GPP [21]. A new fast response mSON at CU is proposed to utilize the F1 interface and control DUs.

SON functionalities at base station require fast feedback. They are dependent on real-time feedback from devices, such as RRC connection messages between eNB and UE, radio conditions reported by UE. The feedback could be handover messages in tens of millisecond interval, or Channel Quality Indicator (CQI) at TTI level. Major self-optimization functions include Automatic Neighbor Relations (ANR), Mobility Robustness Optimization (MRO), Mobile Load Balancing (MLB), and Coverage and Capacity Optimization (CCO). The ANR is defined in 3GPP TS32.511 [28], with others in TS32.521 [29].

cSON functionalities are functions that would require cluster level information. Those would need coordination and relationship among cell sites in a large time interval. Performance Management typically pulls data in a 15 min interval, while call trace in a 1 min interval. Centralized functions will take into account cluster information to optimize the parameter from an overall network perceptive. cSON can further improve eNB functions, and expanded optimization functions at cluster level, such as load balancing among cells and antenna tilting optimization. The cSON can also utilize end-to-end latency measurement to improve Voice over LTE (VoLTE) coverage by dynamically adjusting the number of HARQ retransmissions [31].

Both eNB SON and cSON are needed to ensure proper performance of a network. For example, ANR at eNB can automatically add new sites, and avoid call drops. Centralized ANR will oversee the eNB ANR functions, automatically remove non-used neighbors, or add must have neighbors. One use case is a distance neighbor overshadowing a nearby neighbor and being wrongfully detected by UE. This could end up with ping-pong between source and target cell sites, or cause call drops. The centralized function can remove or deny this neighbor from the neighbor relationship, updating the blacklist to prevent this happening. CCO optimize cell site transmit power, antenna tilting, and others to relieve congestion and fill coverage holes. Those changes could also impact ANR, MRO, and MLB at both cSON and dSON. Thus, the feature coordination and KPI safeguard are important to avoid any unnecessary over-optimization.

The mSON is suitable for local breakout to enable low latency applications, as well as spectrum sharing under sub-second level. It is complementary to the current dSON at the base stations and the cSON functionality at NMS level. It can conduct traffic steering, QoS provisioning, dual connectivity, and spectrum sharing at a finer granularity, while enable handover and load balancing at a cluster level.

The mSON can take full advantage of the F1 interface to facilitate spectrum sharing, carrier aggregation, and dual connectivity among various frequency carriers as illustrated in Figure 2.10a. 5G expands the frequency bands beyond traditional low-bands and mid-bands. The challenging with fast inter-frequency MRO at eNB and a much longer term load balancing among the same carriers at cSON becomes more sophisticated in the 5G network. Therefore, mSON will be in a unique position to enable a fast MRO and MLB, not only from an individual cell and neighbors but also from a cluster point of view. As 5G deployments start on high-band for increased capability and ultra-low latency, dual access links are expected to combine with low and mid-band for increased wide-area coverage. When 5G is maturing, high-band combining with mid-band and low-band can maximize coverage, capacity, and cell edge performance.

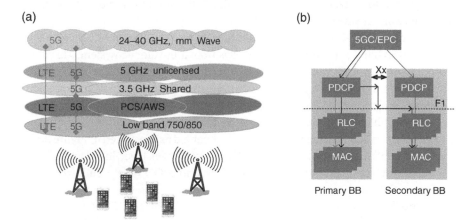

Figure 2.10 Spectrum sharing and multiple links.

Various access options for combining access links for spectrum sharing are illustrated in Figure 2.10b. They can be break-before-make, and only maintain a single link on a spectrum at a time. It can also use eLTE/EPC as the primary link and 5G as the secondary link in 5G early deployment NSA mode, and then using NR/5GC as primary link and eLTE as the secondary link to preserve the LTE investment in SA mode. As 5G-NR is growing, both NR and NR access links can be used for maximal capacity. The Xx interface is referred to as X2 for NSA, and Xn for SA, respectively. The combination can be either dual connectivity among non-collocated cells, or carrier aggregation on collocated cells. Shared and unlicensed spectrum will be utilized to further improve user experience.

As the mSON is collocated with CU, it can coordinate radio and transport resource to reduce any potential congestion, and optimize access and backhaul routing. Being close to where access is happening, it can optimize network slicing for a specific customer to achieve desired user experiences. The mSON can work together with cSON to control traffic routing either locally or through the cloud based on latency, reliability, and other service requirements. It can work closely with dSON to determine the level of power boosting and retransmission or repetition to meet the latency and reliability requirements.

mSON can coordinate well with cSON on network orchestration and dSON on dynamic real-time optimization. mSON is going to play an important role in utilizing resources more efficiently as spectrum and access links increase.

2.3.2 An Intelligent Hybrid 3-Tier SON

By combining traditional SON with mSON, the scope of SON functionalities is expanded. In particular, with the introduction of the DU-CU splitting interfaces at both high and low layers; an intelligent 3-tier SON can autonomously optimize the 5G network at various layers.

Lower-layer split defined a standardized interface among DU and RU, therefore, it opens the possibility to control distributed RUs from a broader ecosystem. This low-latency radio coordination will maximize spectral efficiency and mobility performance among multiple transmission points. It will enable coordinated multiple transmission point features across entire RAN, even for RUs from different manufacturers.

With the addition of lower-layer split, a fully coordinated radio resource management at three different latency levels can be enabled as illustrated in Figure 2.11. The dSON functionality can operate at 250 μs over lower-layer interface, mSON functionality at less than 50 ms over F1 interface, and cSON functionality at above 1 s over north-bound network management interface.

This intelligent hybrid 3-tier SON expands traditional eNB SON function to multiple vendors through standardized lower-layer splitting for provisioning, optimizing, and healing among RUs. It introduces mSON functions for tightly

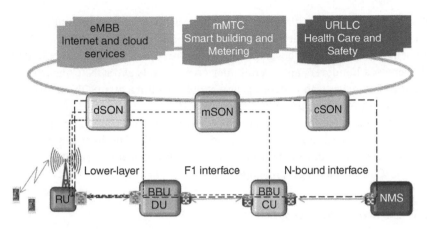

Figure 2.11 Intelligent hybrid 3-tier SON.

coordinated dual connectivity and spectral sharing, enabling traffic steering for low latency application among various digital units. It refines the cSON for overall RAN efficiency and performance.

Operators will define policy with the relative weights for each cell in a centralized database. Those weights could be a combination of RF characteristics and services. The RF characteristics include traditional path-loss, received signal strength or signal-to-interference-plus-noise ratio, network loading characteristic and inter-frequency path-loss discrepancies, operator biasing, and restricted and default values. The service weights could be relative latency, throughput, reliability requirement, and user QoS levels. Several algorithms could also try to optimize the same parameter. The final adjustment could be a combined decision at various levels, and limited to a preset boundary. Those boundaries could be the hard or soft limitation for a particular parameter, e.g. maximal tilting at a RET antenna or a handover parameter. The final changes at eNB could be based on historic and current proposed changes from the distributed, middle and centralized algorithms:

$$P(t) = Pd(t) + g\{Pc(t), Pm(t), Pd(t-1)\}$$

where $P(t)$ is the eNB parameter adjustment. $Pc(t)$, $Pm(t)$, and $Pd(t)$ are the proposed change by centralized, middle-tier algorithm, and eNB algorithm at time t, respectively.

The dynamic control mechanism for multiple layer self-optimizations can execute each layer in an iterative manner triggered by a defined number of events or time periods needed for each layer of optimizing algorithm. Each layer defines operation boundaries of lower layers, a set of eNB parameters and value ranges to optimize upon. A convergent optimized network parameter value is iteratively achieved.

The 3-tier SON is essential to support services with various reliabilities and latencies. A higher reliability application demands higher transmission fidelity leading to a more confined coverage [5]. Therefore, it changes the cell boundary with increased handovers and overall interference levels at dSON. Concurrently, it requires intensive traffic steering and mobile routing at mSON, as well as increased end-to-end resource allocations at cSON.

A unified radio architecture supports programmable radio access components, transport components, and core and cloud components. The hybrid 3-tier SON will allow intelligent control of resources for various applications enabling autonomous MEC at various edges and core networks.

2.3.3 Service-Aware Access Scheme

Operator and vertical provider play an essential and more active role to ensure smooth and coordinated operations of various SON features among a large number of configurable components. An intelligent autonomous network requires a coordinated control system at various layers, among RUs, CUs and DUs, various core network components and NMSs. Operators define a centralized policy management and SON orchestration that include targeted key network and service performance metrics and boundaries of various parameters for various services.

The level of benefits depends on interaction between human and machine insights. Automation is as important as optimization in a commercial operational network. 5G architecture and DU-CU splitting provided enablers to support configurable access schemes based on end-to-end application characteristics and service requirement.

Since QoS profile is available in 5G radio access nodes, the selection of dual connection among 5G-NR and enhanced LTE (eLTE) could include user applications and service requirements, on top of coverage and radio KPIs. This will improve the user quality of experience (QoE) with most efficient uses of network radio and hardware resources.

The mSON mechanism can control the aggregation point of dual connectivity among 5G NR and eLTE. This control mechanism can be adjusted for different network slicing. It can detect device traffic type from transmitted data pattern through deep learning. It maps radio bearers to various packet data sessions with differentiated treatment based on QFI/5QI.

The mechanism can activate various access technologies based on location and radio access coverage availability. For example, it redirects traffic to mitigate the blockage of mmWave or by using eLTE for coverage challenging areas. The device and service profile will inform the device feature and service capability.

A latency tolerant large data transmission will be switched at core to reduce resource consumption. A latency sensitive transmission will be dual connected and aggregated at co-located distributed BBUs, e.g. high-fidelity video. The IMS-based service, or voice service will be directed to eLTE only initially as voice service is not supported in initial NSA specification. The flow chart for this mechanism is illustrated in Figure 2.8.

This processing is continuously performed for all data transmission on an ms interval to maximize performance and efficiency. It activates the selected type of dual connectivity based on coverage, delay requirement, and traffic pattern.

The mechanism in Figure 2.12 minimizes radio and hardware resources while maximizes user performance. The dual connectivity is only enabled for applications requiring large data throughput with low latency at initial 5G launch.

The service-aware access scheme can maximize radio resource utilization and hardware efficiency by analyzing a large amount of network performance data to meet the service requirement. It can reduce unnecessary handover and signaling, consequently improving user experience.

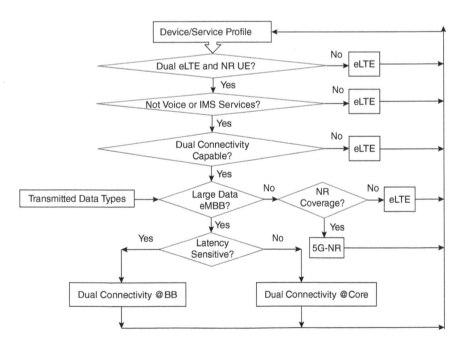

Figure 2.12 Flow chart of application driven 5G-NR and eLTE dual connectivity.

2.3.4 Performance Benefits

An intelligent autonomous network can adjust access means according to network dynamic user and service environment. This improves overall performance. Massive MIMO, an important component of 5G, is used to illustrate the performance benefits.

As network and device increase the number of transmitting and receiving antennas, cell configuration and beam pattern for each site and each user needs to be optimized as well.

The low-layer splitting makes it possible for centralized baseband and dSON to control the RUs from various manufacturers. The beam shaping could be controlled through the m-plane defined in xRAN or low-layer splitting in 3GPP. It can optimize the beam shaping based on cell site configurations and RF characteristics. xRAN has defined digital beamforming in terms of beam Indices, Real-time weights and UE channel information through one or multiple profile files [32,33]. Those files can be modified by SON mechanism through NetConf/Yang model. 3GPP has also completed its study report on SON for an Active Antenna System [34].

RU with advanced massive MIMO capability can change the angle and width of the main lobe, as well as relative power ratio and suppression to various side lobes based on control parameters passing though low-layer interface from digital and central units. There are multiple beam patterns associated with massive MIMO, such as beam patterns defined in 3GPP Release 10 for Common Reference Signal (CRS) shown in Figure 2.13. SON can optimize the network performance by selecting the right beam pattern. For example, a vertical user dominant downtown skyscape would be more efficient from the vertical top beams in Figure 2.13, while

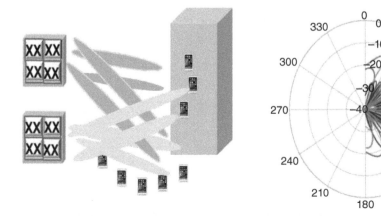

Figure 2.13 Antenna beam patterns.

Table 2.2 Performance optimization from various beam patterns.

	BP#1	BP#2	BP#3	BP#4	Optimized
Spectral efficiency (%)	13	19	22	27	27
Average UE data rate (%)	111	154	176	165	176
UE rate <1 Mbps (%)	−60	−70	−69	−69	−70
PDCCH CCE utilization (%)	−2	−18	−22	−25	−25
DL RLC latency (%)	−46	−56	−55	−55	−56

a hot spot could gain more capacity from a bottom sharp beam created in the right-hand side antenna azimuth pattern.

SON performance benefit is illustrated in Table 2.2 from field results in a commercial trial network. The optimized beam pattern could double the spectral efficiency and average user data rate as compared with default beam pattern for existing legacy devices. Control channel utilization can also be reduced by 20%. In particular, the percentage of users getting low throughput is reduced dramatically. The optimized network configuration also depends on the objective of operator or vertical partners. Beam pattern 4 (BP#4) could provide the best cell capacity, while beam pattern 2 (BP#2) could provide the least number of users with data rate less than 1 Mbps.

Intelligent autonomous operations relieve engineers from a large amount of data, and the complexity of physical and virtual NFs. They can significantly optimize the user and network performance, increase the capacity and enhance the connectivity.

The service-aware SON provides vertical opportunities to define required key performance metrics and optimization criteria for network operation and management. The network operations and SON will act differently for specific vertical applications to maximize benefits for verticals.

2.4 Summary

5G wireless networks have fundamentally evolved from communications between humans to an intelligent service delivery platform that can connect everything. The 5G flat network architecture and standardized high-layer RAN split F1 interface allow more flexible service delivery and a broader ecosystem. An intelligent mobile edge network with a hybrid 3-tier SON will enable autonomous engineering and operation. This is essential to support less than millisecond ultra-low latency applications at 99.999% reliability, on top of eMBB and mMTC.

Verticals can classify their use cases into a combination of those three fundamental 5G service categories to speed up their implementation. Virtualized end-to-end network functionalities will provide opportunities for operators and verticals to define a distinct and isolated logical network of individual use cases in real-time, built upon a shared physically wide-deployed commercial infrastructure. This can be an ultra-reliable low latency service at the far edge for superior reliability and performance. Alternatively, it can leverage the wide-area 5G NR eMBB service for a cost effective way for connections of everything. The 5G network can be readily configured as a private low-cost and high-performance network for verticals. Thus it provides enormous business opportunities for operators and verticals.

The service-aware SON will further provide verticals opportunities to define key performance metrics and optimization criteria for network operation and management. Thus, the mobile wireless network is transforming into an intelligent multiple-purpose autonomous 5G network for a variety of vertical services.

Acronyms

5G	Fifth generation
5GC	5G core
5G-NR	5G New Radio
5QI	5G QoS Indicator
AF	Application Function
AMF	Access and Mobility Management Function
ANR	Automatic Neighbor Relation
AR	Augmented Reality
ARP	Allocation and Retention Priority
BBU	Baseband Unit
CCO	Coverage and Capacity Optimization
CP	Control plane
CPRI	Common Public Radio Interface
CQI	Channel Quality Indicator
CRAN	Centralized RAN
CRS	Common Reference Signal
cSON	Centralized SON
CU	Central Unit
DL	Downlink
DN	Data Network
DNN	Data Network Name
DRB	Data Radio Bearer
dSON	Distributed SON

DU	Distributed Unit
EBH	Ethernet Backhaul
eLTE	Enhanced LTE
EPC	Evolved Packet Core
ETSI	European Telecommunications Standards Institute
FDD	Frequency Division Duplex
gNB	Next generation NodeB
GW-C	EPC Gateway control plane function
GW-U	EPC Gateway user plane function
IMS	IP Multimedia Subsystem
IoT	Internet of Things
KPI	Key performance indicator
LDPC	Low-density parity-check
LTE	Long-Term Evolution
LTE-M	LTE Cat-M1 or Long Term Evolution (4G), category M1
MAC	Medium Access Control
MEC	Multi-access edge computing
ML	Machine Learning
MLB	Mobility Load Balancing
mMTC	Massive Machine Type Communications
MRO	Mobility Robustness Optimization
mSON	Middle-tier SON
NB-IoT	Narrowband Internet of Things
NEF	Network Exposure Function
NF	Network Function
NFV	Network Function Virtualization
NG-C	Next Generation Core, control plane
NG-U	Next Generation Core, user plane
NR	New Radio
NRF	Network Repository Function
NR-IoT	New Radio-Internet of Things
NSSAI	Network Slice Selection Assistance Information
NSSF	Network Slice Selection Function
NSSP	Network Slice Selection Policy
PCF	Policy Control Function
PDCP	Packet Data Convergence Protocol
QFI	QoS Flow Identifier
QoE	Quality of experience
QoS	Quality of service
(R)AN	(Radio) Access Network
RLC	Radio Link Control

RRC	Radio Resource Control
RU	Radio Unit
SA NR	Standalone New Radio
SCS	Subcarrier Spacing
SD	Slice Differentiator
SDN	Software-defined networking
SMF	Session Management Function
S-NSSAI	Single Network Slice Selection Assistance Information
SON	Self-Organizing Network
SSC	Session and Service Continuity
SST	Slice/Service Type
TDD	Time Division Duplex
TTI	Transmission Time Interval
UDM	Unified Data Management
UE	User equipment
UL	Uplink
UP	User plane
UPF	User Plane Function
URLLC	Ultra-reliable and low-latency communications
V2X	Vehicle-to-everything
VR	Virtual reality

References

1 Signals Research Group (2018). 5G: the greatest show on Earth, Vol. 14, No. 8.

2 3GPP TS 38.201-36.215 (2017). 3GPPP, Technical Specification Group, Radio Access Networks; NR; Physical Layers; Release 15, December 2017.

3 3GPP TS23.501-503 (2018). System Architecture: Procedures: Policy and Charging Control Framework for the 5G System, V15.3.0, 2018-09.

4 Kekki, S., Featherstone, W., Fang, Y. et al. (2018). MEC in 5G networks. http://www.etsi.orgimages/files/ETSIWhitePapers/etsi_wp28_mec_in_5G_FINAL.pdf.

5 Sachs, J., Andersson, L., Araújo, J. et al. (2018). Adaptive 5G low-latency communication for tactile Internet services. *Proceedings of the IEEE* (October 2018).

6 Schulz, P., Matthe, M., Klessig, H. et al. (2017). Latency critical IoT applications in 5G: perspective on the design of radio interface and network architecture. *IEEE Communications Magazine* 55 (2): 70–78.

7 Mwanje, S., Decarreau, G., Mannweiler, C. et al. (2016). Network management automation in 5G: Challenges and opportunities. 2016 IEEE 27th Annual International Symposium on Personal, Indoor, and Mobile Radio Communications (PIMRC).

8 Yang, J. and Chan, Y. (2019). Transforming towards an intelligent multiple purpose 5G network. *IEEE Communications Magazine* 14 (2): 53–60.

9 Moysena, J. and Giupponi, L. (2018). From 4G to 5G: self-organized network management meets machine learning. *Computer Communications* 129: 248–268.

10 Klaine, P.V., Imran, M.A., Onireti, O., and Souza, R.D. (2017). A survey of machine learning techniques applied to self-organizing cellular networks. *IEEE Communication Surveys and Tutorials* 19 (4): 2392–2431.

11 Y. Ouyang; Z. Li; L. Suet al. Application behaviors Driven Self-Organizing Network (SON) for 4G LTE networks," *IEEE Transactions on Network Science and Engineering*, Year: 2018 1

12 Polese, M., Jana, R., Kounev, V. et al. (2018). Machine Learning at the Edge: A Data-Driven Architecture with Applications to 5G Cellular Networks, arXiv:1808.07647v1 [cs.NI], August 2018.

13 Arslan, M., Sundaresan, K., and Rangarajan, S. (2015). Software-defined networking in cellular radio access networks: potential and challenges. *IEEE Communications Magazine* 53 (1): 150–156.

14 Mountaser, G., Condoluci, M., Mahmoodi, T. et al. (2017). Cloud-RAN in support of URLLC. IEEE Globecom Workshops.

15 3GPP TR 23.799 (2016). Study on Architecture for Next Generation System (Release 14), V2.0.0, November 2016.

16 Yang, J. (2018). 5G Wireless. In: *Encyclopedia of Wireless Networks*. Springer.

17 Verizon (2017). State of the IoT Market – 2017. www.verizon.com.

18 Yang, J., Song, L., and Koeppe, A. (2016). LTE field performance for IoT applications. *Proceedings of VTC* (18–21 September 2016). IEEE.

19 3GPP TR38.801 (2017). Study on New Radio Access Technology; Radio Access Architecture and Interfaces (Release 14), V2.0.0, March 2017.

20 3GPP TS38.401 (2018). NG RAN, Architecture Description (Release 15), V15.3.0, 2018-09.

21 3GPP TS38.470-474 (2018). NG RAN: F1 General, F1 Layer1, F1 Signaling, F1AP, F1 Data Transport (Release 15), V15.3.0, 2018-09.

22 3GPP TR38.816 (2017). Study on CU-DU Lower Layer Split for NR; (Release 15), V1.0.0, December 2017.

23 3GPP TS 28.532 (2019). Management and Orchestration; Generic Management Services.

24 3GPP TS 28.550 (2019). Management and Orchestration; Performance Assurance.

25 3GPP TS 32.421 (2019). Telecommunication Management; Subscriber and Equipment Trace; Trace Concepts and Requirements.

26 Kakadia, D., Yang, J., and Gilgur, A. (2017). *Network Performance and Fault Analytics for LTE Wireless Service Providers*. Springer.

27 3GPP TS 32.501 (2018). Telecommunication Management; Self-Configuration of Network Elements; Concepts and Integration Reference Point (IRP) Requirements.

28 3GPP TS 32.511 (2018). Telecommunication Management; Automatic Neighbour Relation (ANR) Management; Concepts and Requirements,V15.0.0, 2018-06.

29 3GPP TS 32.521 (2012). Telecommunication Management; Self-Organizing Networks (SON) Policy Network Resource Model (NRM) Integration Reference Point (IRP); Requirements.

30 3GPP TS 32.541 (2018). Telecommunications Management; Self-Organizing Networks (SON); Self-healing Concepts and Requirements.

31 Yang, J., Liu, A., Elmishad, K. et al. (2018).Dynamic HARQ optimization for Voice over LTE. *Proceedings of ICC 2018* (May 2018).

32 xRAN Forum (2018).Control, User and Synchronization Plane Specification. http://www.xran.org/resources.

33 xRAN Fronthaul Working Group (2018). Management Plane Specification, XRAN-FH.MP.0-v01.00, July 2018.

34 3GPP TR32.865 (2017). Study on OAM Aspects of SON for Active Antenna System (AAS) Based Deployments (Release 15).

Part III

5G Verticals – Radio Access Technologies

3

NR Radio Interface for 5G Verticals

Amitava Ghosh, Rapeepat Ratasuk, and Frederick Vook

Nokia Bell Labs, Naperville, IL, USA

Abstract

This chapter discusses various 5G New Radio (NR) features and how these features are used for various industries particularly Industrial Internet of Things (IIoT), vehicle-to-everything (V2X), and eHealth. It first gives an overview of 5G NR radio interface covering numerology, massive multiple-input multiple-output features for both sub-6GHz and millimeter wave spectrum, advanced channel coding followed by the three pillars of NR, namely enhanced mobile broadband, ultra-reliable and low-latency communications, and massive machine type communication. The chapter discusses selected examples of 5G verticals along with how 5G NR features are used effectively in these industries. It shows various use cases for different verticals and how they can be mapped to the three 5G use cases. The chapter also gives a technical overview of three vertical industries, IIoT, automotive V2X, and eHealth, and how 5G NR features are applied to these verticals.

Keywords *5G NR radio interface; 5G verticals; eHealth; enhanced mobile broadband; Industrial Internet of Things; massive machine type communication; ultra-reliable and low-latency communications; vehicle-to-everything*

3.1   Introduction

3GPP NR (Third Generation Partnership Project New Radio) provides a unified, flexible radio interface capable of supporting various 5G verticals such as automotive, healthcare, industry, smart city, utility, smart home, etc. This chapter discusses various 5G NR features and how these features are used for various industries particularly Industrial Internet of Things (IIoT), vehicle-to-everything (V2X), and eHealth. The chapter is organized as follows: Section 3.2 gives an overview of 5G NR radio interface covering numerology, massive multiple-input multiple-output (MIMO) features for both sub-6 GHz and millimeter wave (mmWave) spectrum, advanced channel coding followed by the three pillars of NR, namely enhanced mobile broadband (eMBB), ultra-reliable and low-latency communications (URLLC), and massive machine type communication (mMTC). Finally, Section 3.3 gives a technical overview of three vertical industries, IIoT, automotive V2X, and eHealth, and how 5G NR features are applied to these verticals.

3.2 NR Radio Interface

NR supports three important use cases – eMBB, URLLC, and mMTC [1, 2]. NR is currently supported in two frequency ranges – FR1 from 410 MHz to 7.125 GHz and FR2 from 24.25 to 52.6 GHz. Support for carrier frequency higher than 52.6 GHz will be introduced in the next phase. The underlying technology for the NR radio interface is described in this section. Basic frame structure and numerology are then described in Section 3.2.1 for eMBB (typically the most common NR use case). Enhancements to NR to support URLLC and mMTC use cases are then described in Sections 3.2.2 and 3.2.3, respectively. Please refer to [3] for a detailed overview of the NR physical layer and performance.

With respect to the three use cases, their requirements are shown in Table 3.1.

The NR radio interface provides a flexible framework that can be used to support different use cases as shown in Figure 3.1. This is accomplished through scalable Orthogonal Frequency Division Multiplexing (OFDM)-based numerology and flexible frame structure. Scalable OFDM numerology supports deployment in different spectrum (e.g. low-band, AWS, PCS, mmWave), slot size (from 0.125 to 1 ms), and system bandwidth (from 5 to 400 MHz per carrier). Flexible frame structure allows different services to be offered in the same carrier. For instance, mMTC can be offered at the same time as URLLC by supporting different slot sizes. In addition, NR offers lean carrier design with no cell-specific reference signal and flexible cell-specific signaling such as synchronization signals and broadcast channel. Network slicing allows multiple virtual or logical networks with different end-to-end performance characteristics to be supported using the same physical infrastructure. Some of the performance attributes that can be configured per network slice include latency, throughput, reliability, capacity, mobility, security, analytic, and cost profile. This makes it very easy to offer different services within one NR carrier.

Table 3.1 NR requirements for eMBB, URLLC, and mMTC [2].

	eMBB	URLLC	mMTC
Data rates	20 Gbps for downlink, 10 Gbps for uplink, minimum 100 Mbps downlink and 50 Mbps uplink	N/A	160 bps minimum
User plane latency	4 ms	1 ms	10 s
Control plane latency	10 ms	10 ms	N/A
Reliability	N/A	$1e10^{-5}$ for data packet of size 32 bytes with a user plane latency of 1 ms	N/A
Mobility interruption time	0 ms	0 ms	N/A
System spectral efficiency/ capacity	Deployment dependent – e.g. for dense urban – average spectral efficiency of 7.8 bit/s/Hz downlink and 5.4 bit/s/Hz uplink	N/A	1 000 000 devices per km^2
Coverage	N/A	N/A	164 dB
Battery life	N/A	N/A	10 years

N/A, not applicable.

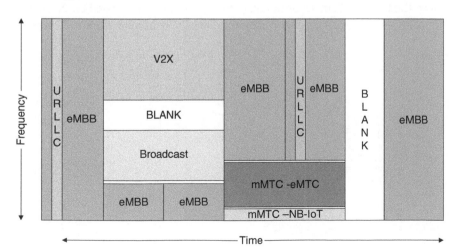

Figure 3.1 Flexible NR framework.

In addition, NR offers the following key radio enhancements:

Massive MIMO and beamforming for higher capacity and coverage

Massive MIMO is the extension of traditional MIMO technology to antenna arrays having a large number of controllable antennas (e.g. 64 and above antenna elements). In the physical layer, the signals from the base station antennas are adjusted via gain or phase control to provide high order spatial multiplexing for enhanced capacity or high gain adaptive beamforming for enhanced coverage. High order spatial multiplexing is used for interference-limited systems (typically those deployed in 6 GHz or below carrier frequency [FR1]), while high gain adaptive beamforming is used for coverage limited systems (typically those deployed in above 6 GHz carrier frequency [FR2]). The 5G NR massive MIMO technology supports both Frequency Division Duplex (FDD) and Time Division Duplex (TDD), i.e. it is duplexing agnostic.

The NR massive MIMO supports scalable and flexible implementation. For sub 6 GHz (FR1) the gNB (i.e. base station) supports full digital array architectures while hybrid/analog architectures are supported above 6 GHz [4]. In NR, all the channels including common channels (e.g. BCH, Synch Signals, RACH) can be beam formed. Beam management is a new set of procedures to assist both the base station and the user equipment (UE) to set their RX and TX beams for both downlink and uplink transmissions and is used both for initial access and connected mode. Beam management supports a hierarchical scan procedure termed a P1 and P2 procedure as illustrated in Figure 3.2. **P-1** uses wide beams for System Synchronization Block (SSB) sweep to scan all directions in a sector and **P-2** then uses transmit Channel State Information Reference Signals (CSI-RS) over narrow beams within the SSB identified by the UE. The **P-3** procedure is used for beam refinement at the UE, which enables the UE to select its best beam for TX and RX.

Next, the Channel State Information (CSI) framework is discussed for NR massive MIMO operation. The major components of CSI framework are (i) report

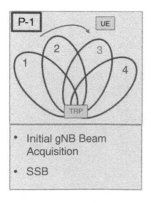
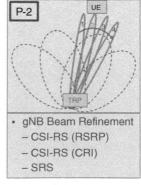
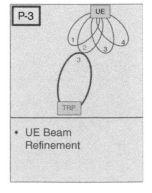

Figure 3.2 NR beam management procedure.

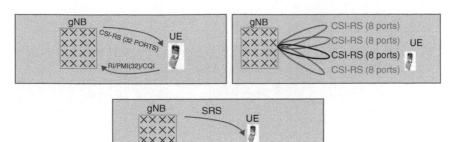

Figure 3.3 CSI-RS modes.

settings, i.e. what CSI to report and when to report it, (ii) resource settings, i.e. what signals are used to compute CSI, and (iii) trigger states which associate (i) and (ii). In the NR framework, the CSI-RS are also used for mobility management, beam management, and tracking in addition to CSI measurements. The downlink NR MIMO operation may utilize single CSI-RS or multiple CSI-RS or Sounding Reference Signal (SRS) for TDD operation. The single CSI-RS may be transmitted over a maximum of 32 ports, and UE computes Rank Indicator (RI)/Precoding Matrix Indicator (PMI)/Channel Quality Indicator (CQI) from the transmitted CSI-RS. Multiple CSI-RS combine beam selection with codebook feedback with gNB transmitting one or more CSI-RS, each in different "directions" (maximum of 8 ports per beam for a total of 32 ports) as shown in Figure 3.3. The UE then computes CSI-RS Resource Indicator (CRI)/PMI/CQI from these CSI-RS. Finally, SRS based operation wherein the UE transmits SRS signal is also supported in NR and exploits TDD reciprocity. Based on the SRS the gNB computes transmit weights used for DL beamforming. NR also supports two types of codebook namely Type-I and Type-II codebook. The Type-I codebook supports standard resolution CSI feedback and can be applied to both single panel and multi-panel antenna array configurations and is primarily aimed at single-user multiple-input multiple-output (SU-MIMO) transmission. Type-II codebook supports high resolution CSI feedback targeting multi-user multiple-input multiple-output (MU-MIMO), and the UL overhead due to high resolution CSI feedback is approximately two to five times that of standard resolution CSI feedback depending on the configuration of the CSI feedback.

Next, the system performance of 5G NR and Long-Term Evolution (LTE) at 2 GHz is summarized with different number of antenna ports and feedback modes. Figure 3.4 shows the antenna configurations for different numbers of antenna ports, both for the physical antenna configuration and the logical transceiver configuration after antenna ports are mapped to transceivers. Figure 3.5 illustrates the sector and edge spectral efficiency (SE) for Urban Macro-cell (UMa) scenario and

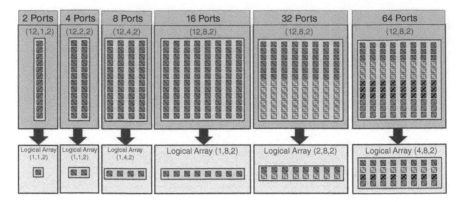

Figure 3.4 Antenna configurations with different numbers of antenna ports.

full buffer traffic with 500 m inter-site distance (ISD) and using single CSI-RS option as defined earlier. The performance in Figure 3.5 is shown for "practical" configurations that are relatively simple to implement and deploy quickly and also for "optimized" implementations that utilize advanced feedback strategies and advanced MU-MIMO precoding based on zero-forcing-style transmission.

The following assumptions were used:

- 2, 4, 8 ports = LTE: 8-port "Practical" = SU-MIMO with Rel-10 CB; 8-port "Optimized" = MU-MIMO with Rel-10 CB with Zero Forcing (ZF).
- 16, 32 ports = NR: "Practical" = MU-MIMO with Type I CB without ZF; "Optimized" = MU-MIMO with Type II CB with ZF.

The following observations are made:

- There is ~2–3 times gain in going from the LTE 8-port practical system to the 16- and 32-port NR optimized systems, respectively.
- The optimized system uses zero forcing transmitter, MU-MIMO and Type-II feedback.

Usage of mmWave spectrum

The first phase of NR supports deployment in carrier frequency up to 52.6 GHz, while later phase will support deployment beyond 52.6 GHz. Deployment in the mmWave band is very attractive due to the large amount of available spectrum. For instance, spectrum blocks on the order of 1–2 GHz are available in the mmWave band, while only 100 MHz or so are available in the AWS band.

mmWave systems must overcome high path loss and diminished diffraction. To overcome high path loss and potential blockage in the mmWave spectrum, NR has been designed to naturally support beam-based operation using antenna arrays with large number of antenna elements for both control and data channels

(a)

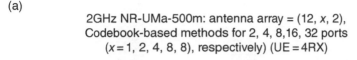

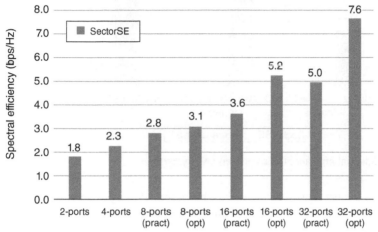

(b)

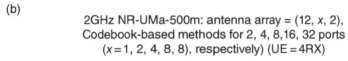

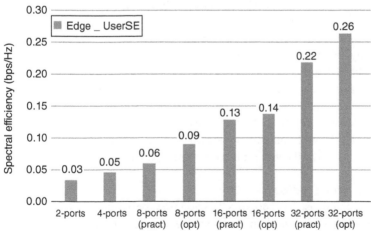

Figure 3.5 (a) Sector and (b) edge spectral efficiency for LTE and NR.

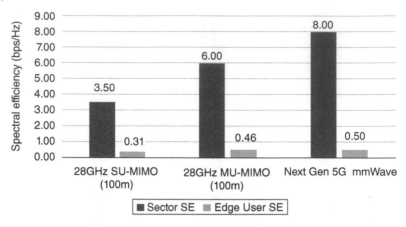

Figure 3.6 Spectral efficiency trends of mmWave systems.

as outlined in the previous section [4]. Furthermore, NR supports phase-tracking reference signal which can be used to compensate from phase noise that is more prevalent in higher frequency.

Cell coverage in mmWave is typically small due to high path loss and high probability of blockage. Thus, mmWave cells are provided for capacity enhancement, and densification is needed to achieve 95% coverage reliability. Fiber penetration particularly in North American cities and Europe do not match the ISD requirement of 100–200 m required for deployment of mmWave systems. As such, Integrated Access and Backhaul (IAB) nodes may be deployed for four fundamental reasons: (i) to mitigate sparse fiber; (ii) to remediate isolated coverage gaps; (iii) to enhance capacity; and (iv) to bridge coverage from outdoor to indoor. IAB is currently being standardized in 3GPP Rel-16 [5].

Figure 3.6 shows the potential sector and edge SE for current and next generation mmWave systems at 28 GHz assuming ISD of 100 m, TDD 50 : 50 split, AP with cross pole array of 512 elements and UE with 32 elements cross pole array using both SU-MIMO and MU-MIMO. It may be observed that the SE of mmWave systems with SU-MIMO is similar to NR sub- 6 GHz system and the SE can be enhanced using MU-MIMO with multiple antenna panels.

Advanced channel coding

On the data channels, NR supports Low-density parity-check (LDPC) code, which is defined by parity check matrices with a quasi-cyclic design. This design enables high-throughput and low-latency hardware implementation as the LDPC decoder can be fully parallelized in hardware. In terms of performance, LDPC performs slightly better than the Turbo code used in 4G LTE. It, however, exhibits an error floor at very low Block Error Rate (BLER). This is not crucial for data channels as HARQ retransmission can be used to ensure the packet is decoded correctly. However, HARQ retransmission is not possible in control channels. Therefore, to eliminate this error floor, Polar code was adopted in NR for control channels. In addition to no error floor, Polar code performs better than LDPC for short blocks, which is typical of the payload for control channels [6]. Compared with LDPC, however, Polar code implementation is more complex but still efficient for control channels with short code blocks.

3.2.1 eMBB

The NR radio interface is based on OFDM with scalable numerology to address different spectrum, bandwidth, deployment, and services. The system can be scaled by using different subcarrier spacing (SCS) of 15, 30, 60, and 120 kHz. For example, deployment in mmWave frequency band may use 120 kHz SCS to support very large carrier bandwidth (e.g. 400 MHz) and provide robustness to phase noise. For macro deployment in lower frequency band, 15 kHz SCS may be sufficient since the carrier bandwidth is typically limited to 20 MHz or less and lower SCS allows for more efficient use of smaller spectrum with better support for mobility. In addition, within the same carrier, multiple bandwidth parts, each with potential different SCS can be supported as shown in Figure 3.7. In 5G NR, a bandwidth part is defined

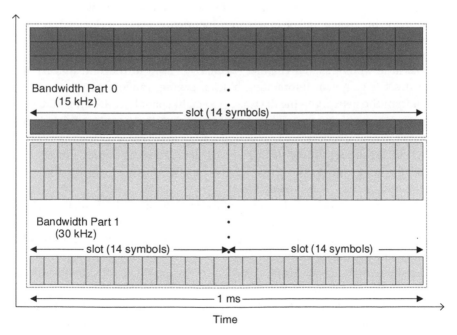

Figure 3.7 NR deployment supporting different bandwidth part numerologies.

as a contiguous set of physical resource blocks (PRBs), selected from a contiguous subset of the common resource blocks for a given numerology. Up to four bandwidth parts can be defined in downlink and uplink, and each bandwidth part can contain its own synchronization signal block. This allows different services with different requirements (e.g. latency) to be deployed in the same NR carrier. In addition, it also supports having UE with smaller bandwidth than the wideband carrier, or UE that may support wideband carrier using multiple RF chains (similar to UE supporting intra-band carrier aggregation in 4G LTE).

In the frequency domain, the spectrum in each bandwidth part is divided into PRBs, each of which is composed of 12 subcarriers. For 15 kHz SCS, a PRB occupies 180 kHz, while for 60 kHz SCS, a PRB will occupy 720 kHz. User can be allocated resources in multiple of PRBs. The maximum number of PRBs per carrier is 275, giving a maximum bandwidth of 400 MHz. Larger bandwidth can be supported using carrier aggregation (e.g. 2 GHz allocation can be supported using aggregation of 5×400 MHz carriers).

In the time domain, each radio frame is 10 ms long and composed of a variable number of slots. Each slot is composed of 14 OFDM symbols. However, the duration of each OFDM symbol depends on the SCS, with shorter duration as the SCS increases (e.g. the symbol duration for 60 kHz is four times shorter than at 15 kHz). For instance, a slot is 1 ms for 15 kHz SCS but 0.25 ms for 60 kHz SCS. In the time domain, user can be allocated a fraction of slot, one slot, or multiple slots.

The following signals and channels are supported.

Downlink:

- Physical downlink shared channel (PDSCH) – used to transmit unicast and broadcast (e.g. system information blocks, paging, random access response) data from the network to the device. Up to eight spatial MIMO layers are supported. Four modulation sizes are supported – QPSK, 16-QAM, 64-QAM, and 256-QAM. LDPC code is used.
- Physical broadcast channel (PBCH) – used to transmit master information block for initial access. The physical downlink control channel (PDCCH) uses LDPC coding and QPSK modulation.
- PDCCH – used to transmit downlink control information (DCI) such as scheduling grant, slot format indication, preemption indication and power control information. Eight different DCI formats are supported. The PDCCH uses Polar coding and QPSK modulation.
- Demodulation Reference Signal (DMRS) – reference signal for channel estimation and demodulation.
- Phase Tracking Reference Signal (PTRS) – reference signal for phase tracking and correction. Phase noise increases with carrier frequency but is generally not an issue below 6 GHz. At mmWave, phase noise can vary up to a few degrees from

one OFDM symbol to the next and therefore tracking of phase noise is needed. In addition, the UE can also use PTRS together with DMRS for demodulation.

- CSI-RS – reference signal for channel state measurement by the device.
- Primary synchronization signal (PSS) – time-frequency synchronization.
- Secondary synchronization signal (SSS) – time-frequency synchronization and cell identification acquisition.

Uplink:

- Physical uplink shared channel (PUSCH) used to transmit unicast data and control information. Both OFDM and an optional SC-FDMA waveforms are supported. Five modulation types are supported – $\pi/2$–BPSK (with transform precoding), QPSK, 16-QAM, 64-QAM, and 256-QAM. LDPC coding scheme is used, and up to four spatial MIMO layers can be supported.
- Physical uplink control channel (PUCCH) – used to transmit control information such as HARQ acknowledgment, channel state information, channel quality report, and scheduling request. Five different uplink control information formats are supported. The PUCCH uses QPSK modulation and different coding schemes based on the information to be transmitted.
- PBCH – used to transmit random access preamble as part of initial access procedure. NR supports four PRACH formats and numerologies for different use cases – normal cell, large cell, coverage enhancement, and high-speed cell. The random access sequence design is based on a Zadoff–Chu sequence and supports two sequence lengths – short and long. Short sequence length has lower complexity and better timing accuracy. However, it has lower capacity and can support smaller maximum cell size. Long sequence length has high capacity and can support cell size up to 100 km. However, it has higher complexity and some performance loss at high speed.
- DMRS – reference signal for channel estimation and demodulation.
- PTRS – reference signal for phase tracking and correction.
- SRS – reference signal for channel state measurement by the network.

NR supports both sector-beam (i.e. wide beam) as well as beam-based (i.e. narrow beam) operation. Beam-based operation can be used with high gain massive antenna array for enhanced coverage. For beam-based operation, additional beam acquisition and refine steps are used as described below.

In NR, beam sweeping is used where, at predefined amounts of time, the same information is being sent sequentially across beams. Figure 3.8 shows initial access procedure with beam-based operation. In this case, each synchronization block (containing synchronization signals and the PBCH) is transmitted across multiple beams sequentially. The UE finds the best beam and decodes the system information on that beam. The gNB also provides information related to system

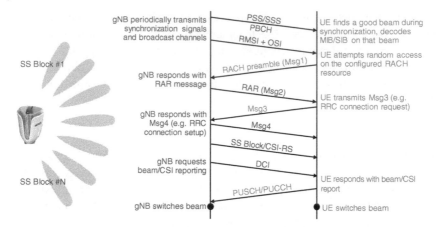

Figure 3.8 Initial access with beam-based operation.

access and random access on the configured random access resource associated with the best beam. UE then performs a four-step random access procedure with the following steps – preamble transmission, random access response, radio resource control (RRC) connection request, and RRC connection complete.

After initial beam acquisition, the gNB can configure CSI resources for further beam measurement and refinement. During this procedure, the gNB can transmit using finer beams to the UE. The UE can respond with the beam report and the gNB can switch the UE to a better beam if appropriate. This process is described in Section 3.2 as part of the massive MIMO operation.

Since the NR system is capable of very high throughput, data service for eMBB may be bursty as user data transmissions can be served very quickly by the network. In Rel-16, a study is being conducted to maximize energy efficiency based on possible traffic profiles. The study aims to identify techniques for UE power saving in connected mode, e.g. wake-up signal to indicate pending data transmission, adaptation of UE on/off period to traffic, reducing the amount of time UE has to monitor the network, and reducing the required number of measurements [7]. Enhancements related to UE power saving are expected to be standardized by the end of 2019.

3.2.2 URLLC

The NR system as described in Section 3.2.1 provides a good foundation for URLLC, which requires very low latency and very high reliability. NR supports both semi-static and dynamic resource sharing between eMBB and URLLC. The gNB can semi-statically allocate separate time-frequency resources to URLLC and eMBB traffic. In addition, dynamic resource allocation is by gNB implementation and gNB can preempt ongoing eMBB traffic for URLLC transmission.

To support user plane latency requirement of 1 ms, the following techniques can be used [8, 9]:

- Short frame structure – short slot length via higher SCS (e.g. slot length at 120 kHz SCS is 0.125 ms), data transmission shorter than a slot (i.e. 2, 4, 7 symbols in downlink, any number of symbols in uplink), and flexible slot structure in TDD (i.e. a slot containing symbols for both data and HARQ feedback).
- Preemption scheduling and multiplexing of URLLC data within eMBB – gNB can preempt (i.e. puncture) on-going data transmission with URLLC transmission.
- Configured grant uplink transmission – UE is assigned preconfigured grant for uplink transmission and does not have to transmit a scheduling request to the gNB. The periodicity of the configured resources can be 2 symbols, 7 symbols, 1 slot, ..., up to 5120 slots.
- Short PUCCH format – PUCCH transmission in one or two symbols to minimize HARQ feedback.
- Short periodicity for scheduling request – UE can be configured with a scheduling request slot as often as every 2 symbols.
- Multi-slot repetition for data channels – allows retransmissions without waiting for HARQ-ACK feedback. It can be configured for 2, 4, and 8 repetitions.
- Shorten UE processing time – NR design facilitates UE pipelining processing to shorten the UE processing time. This is done by transmitting DMRS at the beginning of the slot and using frequency-first mapping which allows symbol-by-symbol processing.

Figure 3.9 illustrates the timeline for downlink transmission, while Figure 3.10 illustrates the timeline for uplink transmission using configured grant transmission.

To support ultra-reliable packet reception, the following techniques can be used:

- Multi-slot repetitions for PDSCH and PUSCH – repetitions to improve data channel reliability.
- High aggregation level for PDCCH – up to 16 aggregation level support for PDCCH.

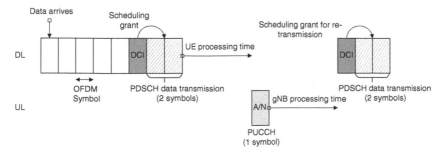

Figure 3.9 URLLC downlink transmission.

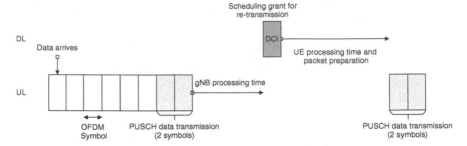

Figure 3.10 URLLC uplink transmission using configured grant operation.

- Enhanced CSI reporting – BLER target of 10^{-5} for CSI reporting, and the corresponding CQI table that includes entries with lower SE.
- PDCP layer data duplication – allows transmission of the same data packet from different transmission points to increase reliability.
- URLLC Modulation and Coding Scheme (MCS) table – additional MCSs to support lower SE targeting toward lower BLER target.

Further URLLC enhancements are being studied in Rel-16. They include downlink control channel enhancements (e.g. compact scheduling grants, additional repetitions, increased monitoring capabilities), uplink control information enhancements (e.g. enhanced HARQ, CSI feedback), uplink data channel enhancements (e.g. mini-slot level hopping), enhancements to scheduling/ HARQ/CSI processing timeline, and enhanced uplink configured grant transmission. In addition, to support IIoT application with time sensitive network, 3GPP is looking at using multi-TRP/panel transmission to improve reliability, and multi-beam operation to improve latency and reliability.

3.2.3 mMTC

mMTC for 5G NR provides low-power wide-area cellular deployment for the Internet of Things (IoT). It is intended to serve low data-rate, delay tolerant traffic where low-cost devices and extended coverage are required. With respect to mMTC, 5G NR will support the following objectives:

- Ultra-low complexity and low-cost IoT devices and networks.
- Maximum Coupling Loss (MCL) of 164 dB for a data rate of 160 bps at the application layer.
- Connection density of 1 million devices per square kilometer in an urban environment.

Table 3.2 eMTC and NB-IoT feature summary [12, 13].

	eMTC	NB-IoT
Carrier bandwidth	1.4, 3, 5, 10, 15, 20 MHz	180 kHz, multiple carriers can be deployed for additional capacity
Device bandwidth	1.4/5/20 MHz	180 kHz
Peak data rates	1.4 MHz device: 1 Mbps downlink, 3 Mbps uplink 5 MHz device: 4 Mbps downlink, 7 Mbps uplink 20 MHz device: 27 Mbps downlink, 7 Mbps uplink	127 kbps downlink, 159 kbps uplink
Duplex mode	Half-duplex/Full-duplex	Half-duplex
Device transmit power	14/20/23 dBm	14/20/23 dBm
Number of antenna at device	1	1
Support for multi-cast transmission	Yes – SC-PTM	Yes – SC-PTM
Support for voice services	Yes – VoLTE	No
Handover support for mobility	Yes	No (only cell reselection supported)
Location services	Yes – OTDOA, ECID	Yes – OTDOA, ECID

ECID, Enhanced Cell ID.

- Battery life in extreme coverage beyond 10 years (15 years is desirable). Battery life is evaluated at 164 dB MCL with mobile originated data transfer consisting of 200 and 50 bytes of uplink and downlink data per day, respectively, and a battery capacity of 5 Wh.
- Latency of 10 s or less on the uplink to deliver a 20-byte application layer packet measured at 164 dB MCL.

In 4G LTE, 3GPP has specified two low-power wide-area technologies for IoT – enhanced MTC (eMTC) [10] and Narrowband IoT (NB-IoT) [11]. eMTC is intended for mid-range IoT applications and can support voice and video services, while NB-IoT can provide very deep coverage and support ultra-low-cost devices. Core specifications for both technologies were completed in 2016. Further work to enhance the technology has been ongoing in 3GPP. Rel-14 and Rel-15 enhancements have been completed, while Rel-16 enhancements are ongoing and are expected to be completed by the end of 2019. Table 3.2 summarizes key features of the two systems.

Evaluations of eMTC and NB-IoT demonstrated that they can satisfy mMTC 5G requirements [14]. Thus, 3GPP agreed not to specify NR-based technology for mMTC, but to reuse eMTC and NB-IoT technologies to support mMTC services. That is, eMTC or NB-IoT can be deployed within NR carrier to support mMTC services. This is also beneficial as it can support legacy LTE IoT devices. It is expected that eMTC and NB-IoT devices will be around for a long time (e.g. water/power meters can have a lifetime of 10–15 years or more). As LTE systems are re-farmed to NR, IoT devices can continue to be supported as part of NR deployment.

3.2.3.1 eMTC Overview

In 3GPP Rel-13, eMTC was introduced. To keep device cost low, a new UE category (Cat-M1 UE) was introduced. Cat-M1 UE has an RF bandwidth of 1.4 MHz (compared with 20 MHz for normal LTE devices), one receive antenna chain (compared with two for normal LTE devices), maximum transport block size of only 1000 bits, and typically operates in half-duplex mode. Despite being bandwidth-limited, Cat-M1 UE can operate within any LTE system bandwidth. This is accomplished by defining special procedures and channels in the cells for bandwidth-limited operation.

Additional features that are supported for eMTC include mobility (i.e. handover), extended discontinuous reception (eDRX), RRC connection suspend/resume, and data transmission via control plane.

In 3GPP Rel-14, key enhancements were introduced including new higher capability UE category (Cat-M2) supporting 5 MHz bandwidth, higher peak data rates for both Cat-M1 and Cat-M2 UE categories, multicast downlink transmission based on single-cell point-to-multipoint (SC-PTM), support for Observed Time Difference of Arrival (OTDOA) positioning, and increased Voice over Long-Term Evolution (VoLTE) coverage for eMTC.

In 3GPP Rel-15, further enhancements were introduced, including reduced system acquisition time by improving cell search and system information acquisition performance, latency and power consumption reduction techniques (wake-up signal or channel, relaxed monitoring for cell reselection, data transmission during random access procedure) and techniques to improve SE such as 64-QAM support, sub-PRB resource allocation, etc.

In 3GPP Rel-16, further enhancements are being introduced for eMTC, including group-based wake-up signal, grant-free uplink transmission, scheduling multiple transport blocks using a single grant, and coexistence with NR.

3.2.3.2 NB-IoT Overview

In 3GPP Rel-13, NB-IoT was introduced. NB-IoT is a NR access system built from existing LTE functionalities with essential simplifications and optimizations. At the physical layer, NB-IoT occupies 180 kHz of spectrum. In the downlink, NB-IoT fully inherits downlink numerology from LTE. The SCS is 15 kHz and 12 subcarriers make up the 180 kHz channel. In the uplink, NB-IoT supports the same LTE numerology but the UE may be assigned 1, 3, 6, or 12 tones. A 3.75 kHz SCS is also supported, with a NB-IoT carrier spanning over 48 subcarriers and occupying 180 kHz as well. For this numerology, four times expansion on the time domain applies to remain compatible with the LTE numerology and the UE is always assigned a single tone.

In the higher layers, simplified LTE network functions are supported. They include idle mode mobility, eDRX, power saving mode, paging, positioning based on the existing location services architecture, and access control. In addition, two optimizations for small data transmission have been specified. The first one is the ability to transmit a small amount of data in the control plane via Signaling Radio Bearer (SRB). The second one is the ability to suspend and resume RRC connection, thus eliminating the need to establish a new connection at each reporting instance.

In 3GPP Rel-14, key enhancements were introduced including new higher capability UE category, multicast downlink transmission based on SC-PTM, and support for OTDOA positioning.

In 3GPP Rel-15, further enhancements were introduced, including reduced system acquisition time by improving cell search and system information acquisition performance, latency and power consumption reduction techniques (wake-up signal

Figure 3.11 mMTC deployment: (a) 20 MHz NR with a 5 MHz eMTC carrier and (b) 20 MHz NR with NB-IoT carriers.

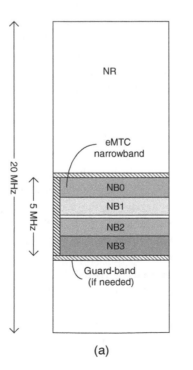

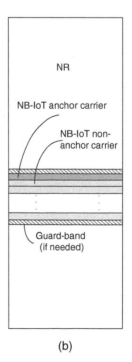

(a)   (b)

or channel, relaxed monitoring for cell reselection, data transmission during random access procedure).

In 3GPP Rel-16, further enhancements are being introduced for eMTC and NB-IoT, including group-based wake-up signal, grant-free uplink transmission, scheduling multiple transport blocks using a single grant, and coexistence with NR.

3.2.3.3 Coexistence with NR

As discussed, mMTC deployment for NR will be supported using eMTC and NB-IoT. In addition, legacy LTE IoT devices will be around for a long time (e.g. water/power meters can have a lifetime of 10–15 years or more). As LTE systems are re-farmed to NR, there would be no need to support broadband LTE UE. However, LTE IoT UE would still need to be supported. Therefore, eMTC and NB-IoT carriers can be deployed within an NR carrier as shown in Figure 3.11. In Figure 3.11a, a 5 MHz eMTC carrier is deployed within a 20 MHz NR carrier. eMTC can support carrier bandwidth of 1.4, 3, 5, 10, 15, and 20 MHz and the size of the eMTC carrier can be chosen based on the IoT load. In Figure 3.11b, multiple NB-IoT carriers are deployed within a 20 MHz NR carrier. Each NB-IoT carrier occupies 180 kHz.

To deploy eMTC within the NR carrier, LTE carrier with eMTC in-band must be used. Since eMTC must follow the 100 kHz raster, its placement within the NR is limited if frequency grid alignment is desired although this is not expected to be an issue. In 3GPP Rel-15, 14 different frequency bands have been specified that can support both NR and eMTC (e.g. Band 8 with uplink from 880 to 915 MHz and downlink from 925 to 960 MHz). So far, only NR bands below 2.7 GHz support eMTC deployment as high-frequency bands are not suitable for IoT services due to limited coverage. Both NR and LTE share a common 100 kHz raster, so the raster grid for the two systems can be aligned. Different LTE bandwidth can be deployed based on expected eMTC load.

To deploy NB-IoT within the NR carrier, standalone or guard-band NB-IoT operation mode can be used. As for eMTC, NB-IoT must follow the 100 kHz raster for the anchor carrier, and therefore its placement within the NR carrier bandwidth is limited if frequency grid alignment is desired although this is not expected to be an issue. In addition, non-anchor carrier can be placed is more flexible locations. In LTE, NB-IoT in-band operation mode was defined to provide coexistence with legacy LTE signals and channels such as CRS, PDCCH, PCFICH, and PHICH. However, as there are no fixed NR signals or channels to be avoided, there is no need to define an in-band operation mode with NR. A reserved time-frequency region, including possible guard if needed, can be used for NB-IoT. NR deployment is very flexible and can be configured via implementation not to overlap with NB-IoT.

Another potential deployment aspect is to deploy NB-IoT using LTE guard-band operation mode within the NR carrier. This may be beneficial in terms of possible placement of NB-IoT carrier with respect to the NR PRB. This is because guard-band operation mode can support frequency offsets to the 100 kHz raster, therefore offering more potential placement of NB-IoT carrier(s) in alignment with NR. As for the case with standalone deployment, we do not see any coexistence issue with placing guard-band NB-IoT carrier(s) within the NR carrier.

In addition, reserved resources as defined in NR can be used to reserve time and frequency resource for deployment of eMTC and IoT. This is done using RB-level bitmap in the frequency domain and symbol-level bitmap in the time domain with repetition pattern. Note that this can be done dynamically in NR and therefore can allow the gNB scheduler to dynamically take advantage of unused eMTC resource. This allows for efficient coexistence of eMTC and NR. For example, in an NR system with 20 MHz bandwidth, less than 5% of the NR resource must be reserved for always-on eMTC and NB-IoT signals and channels such as reference signals, broadcast channels, etc.

In Rel-16, several enhancements are being studied to further improve coexistence performance [15]. They include:

- eMTC and NB-IoT reserve resource. In NR, reserved resource can be used to reserve time and frequency resource for deployment of eMTC. To provide better coexistence with NR, eMTC resource reservation can also be introduced. This is currently under study in 3GPP. Several potential use cases have been considered, including supporting dynamic TDD in NR, avoiding collision with periodic NR transmission, and avoiding collision with URLLC transmission. To support reserved resources in eMTC, a bitmap can be used to mark these subframes as invalid for legacy UEs, while Rel-16 eMTC UE will be allowed to use the available symbols/slots on these subframes. Subframe-level operation can easily be supported with minimal changes. However, symbol-level operation is not defined for eMTC and NB-IoT so this must be introduced. However, it can also be done in a simple manner. For instance, the subframe can follow special subframe configuration defined in LTE which the UE already knows how to handle. Or these symbols may only be used as part of a repetition where symbols from previous subframes are repeated. Alternatively, the subframe can be constructed as before and unavailable symbols can be punctured out.
- User of LTE control region for eMTC. In addition, in Rel-16, standalone deployment for eMTC is being specified. This will enable the use of the LTE control channel region for downlink control and data transmission while still supporting legacy operation for legacy UEs. This can be as simple as repeating or rate-matching downlink transmission in the OFDM symbols that are currently reserved for the LTE control region. Only one or two OFDM symbols would be available, depending on the bandwidth, and the repetition may be as simple as selecting the symbol(s) with the same CRS pattern for repetition. This can be

done for UE in both connected and idle mode as the repetition can always be there even if the UE is not aware of it. For example, the eNB can repeat paging information in the unused symbols. In this case, there would be performance improvement for Rel-16 UE while there would be no impact on legacy UE as it will ignore those symbols.

- Overlap of NR with eMTC and NB-IoT. Generally, eMTC and NB-IoT carrier placement would be done such that overlap with NR is avoided or minimized. This can limit the potential frequency locations that can be used, especially for small NR bandwidth. For example, it is not expected that eMTC will overlap in time-frequency with NR SSB. It is noted that slight modification can be made to NR to support eMTC deployment within the SSB bandwidth. For example, SSB transmission may be arranged to exclude eMTC/NB-IoT symbols.

3.3 5G Verticals

5G NR and LTE are tailored to support various vertical industries addressing three major usage scenarios namely eMBB, URLLC, and massive MTC. In this section, selected examples of 5G verticals are discussed along with how 5G NR features are used effectively in these industries. Table 3.3 shows various use cases for different verticals and how they can be mapped to the three 5G use cases.

In this section, three different verticals – industrial IoT, automotive, and eHealth – will be described together with how 5G can be used to support these verticals.

3.3.1 Industrial IoT

The fourth industrial revolution or the digital enterprise will rebuild the business models in various industries through connectivity. Since different industries have different needs and use cases, connectivity needs to move from a one size fits all to a network that can meet the often specialized and stringent requirements. Industrial IoT networks comprise various industries such as transport venues and ports, warehouse, military bases, power generation, water utility plants, mining, hospitals, and labs to name a few. The drivers for LTE and 5G NR in industrial networks are (i) increased productivity, (ii) digital transformation through wireless technology, and (iii) the use of private networks. Digital transformation in industrial enterprise will utilize the latest 4G and 5G technologies including wireless communication, IoT devices, cloud computing, and analytics and automation tools. A private network is considered ideal for retaining control of the network (network resources and operations view/tools) and security (traffic within

Table 3.3 5G use cases for different verticals.

	Automotive	Infrastructure	Smart home	Smart city	eHealth	Smart factory
eMBB	Assisted driving, Infotainment	Video surveillance	Gaming, remote computing	Video surveillance	Remote diagnostics	Remote diagnostics
mMTC	Remote diagnostics, traffic management	Remote sensors, metering, fleet management, asset tracking	Home sensors, home security	Remote sensors, metering, fleet management, asset tracking, traffic management	Patient tracking, wearables, disaster management, asset tracking	Remote sensors, security, asset tracking
URLLC	Automated driving, public safety	Tele-protection	AR/VR gaming, remote office	Drone, tactile internet	Patient tracking, remote surgery, diagnostics	Industrial IoT, time-sensitive network

enterprise systems). A dedicated 5G/LTE network can provide the throughput, latency, reliability, wide coverage, and mobility requirements of business critical operations while ensuring high security as the convergence of information and operational technologies bring greater susceptibility to cyber attacks. A private 5G network, particularly when leveraging small cells and low, mid and high spectrum bands, allows service customization to fit the industry's needs and isolate threat typical over public networks.

The IIoT network will be scalable in connectivity and number of devices and will be designed to deliver optimal performance for all industrial applications using eMBB, low power IoT, sub-ms delay for latency critical applications, enhanced coverage, accurate positioning and secure connectivity within the enterprise network. In the manufacturing industry, productivity must be improved while also improving the safety. Processes must be automated and production uptime maximized to minimize business interruption to gain more efficiency and competitiveness and value for investments. At the same time, carbon emissions should be reduced, and energy saved. In summary, the technology enablers for an industrial IoT revolution are:

- **5G wireless converged automation protocols:** This will eliminate wires and support time synchronous operations.
- **Private edge cloud:** This feature will be scalable and provide secure local computing.
- **Slicing for IIoT networks:** This critical feature can support multiple stakeholders on one common infrastructure.
- **Machine learning-enabled automated operations**: This will support expertless monitoring, prediction, and optimization.

The performance requirements with respect to latency and reliability for some IIoT use cases are summarized in Table 3.4.

The key 5G NR features which will drive IIoT are URLLC, mMTC, 5G positioning, time-sensitive communication (TSC) and to a lesser degree eMBB. The details of the above features have been described in the previous sections. In Rel-16, native support in 5G for TCS services is currently being studied. The new R16 enablers on top of enhanced URLLC for TSC are [17]:

- Common understanding of global time among devices and network.
- Network to support bounded latencies (both minimum and maximum) and error rates for packet transport.
- More detailed service descriptors for TSC flows (a condition for deterministic forwarding in TSN/DetNet).

Rel-15/Rel-16 URLLC aims for higher reliability and better efficiency and the various features outlined in Figure 3.12, will be able to support various IIoT use cases outlined in Table 3.4 [18].

Table 3.4 URLLC requirements for various IIoT cases [16].

Scenario	End-to-end latency	Communication service availability	Reliability	User experienced data rate	Payload size	Traffic density	Connection density	Service area dimension
Discrete automation – motion control	1 ms	99.9999%	99.9999%	1 Mbps up to 10 Mbps	Small	1 Tbps/km^2	100 000/km^2	100×100×30 m
Discrete automation	10 ms	99.99%	99.99%	10 Mbps	Small to big	1 Tbps/km^2	100 000/km^2	1000×1000×30 m
Process automation – remote control	50 ms	99.9999%	99.9999%	1 Mbps up to 100 Mbps	Small to big	100 Gbps/km^2	1000/km^2	300×300×50 m
Process automation – monitoring	50 ms	99.9%	99.9%	1 Mbps	Small	10 Gbps/km^2	10 000/km^2	300×300×50
Electricity distribution – medium voltage	25 ms	99.9%	99.9%	10 Mbps	Small to big	10 Gbps/km^2	1000/km^2	100 km along power line
Electricity distribution – high voltage	5 ms	99.9999%	99.9999%	10 Mbps	Small	100 Gbps/km^2	1000/km^2	200 km along power line
Intelligent transport systems – infrastructure backhaul	10 ms	99.9999%	99.9999%	10 Mbps	Small to big	10 Gbps/km^2	1000/km^2	2 km along a road
Tactile interaction	0.5 ms	[99.999%]	[99.999%]	[Low]	[Small]	[Low]	[Low]	TBC
Remote control	[5 ms]	[99.999%]	[99.999%]	[From low to 10 Mbps]	[Small to big]	[Low]	[Low]	TBC

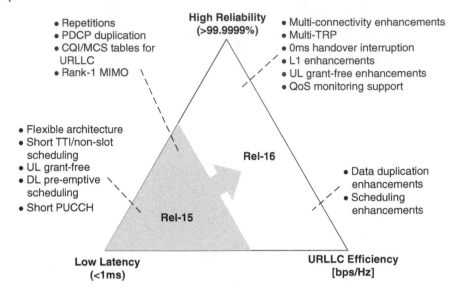

Figure 3.12 Summary of Rel-15/16 features used for IIoT.

NR-based positioning is currently a study item in 3GPP and positioning functionality is expected to be introduced by the end of 2019. NR-based positioning is added to provide added localization capability such as improved accuracy (e.g. cm-level accuracy), utilization of massive antenna arrays, exploitation of spatial and angular domain, and utilization of wide signal bandwidth [19]. In addition to improvement in the physical layer, enhanced positioning architecture is expected to be defined. This enhanced architecture can support hybrid positioning, integrated IoT service, efficient signaling, and reduced end-to-end latency.

In summary, the following 3GPP features will be required for IIoT:

- Enhancements for latency and reliability in radio and e2e (URLLC).
- Support for Wireless Industrial Ethernet and deterministic communications (URLLC).
- 5G private networks in licensed ad unlicensed spectrum (URLLC and eMBB).
- Positioning for 5G and IIoT (eMBB, URLLC, mMTC).
- Cellular IoT evolution, e.g. connecting eMTC/NB-IoT to 5G Core (mMTC).

3.3.2 Automotive V2X

Similar to the Industry 4.0 revolution, the automotive sector is undergoing an important technological transformation as more vehicles are connected to the network. 5G provides wireless technology that can meet the connectivity requirements of the automotive industry. Examples of use cases from the automotive vertical and how they are mapped to the 5G NR features are shown in Figure 3.13.

In 3GPP Rel-15, many use cases for the automotive vertical as shown in [20] can be supported (e.g. fleet management, infotainment, remote diagnostics). However, NR currently does not support direct communication through the sidelink between vehicles and nearby vehicles, infrastructure nodes, or pedestrians (i.e. V2X). Such direct communication can be used to convey information such as position, speed, and heading without having to go through the network. In Rel-16, NR studied advanced V2X services that can be categorized into four use case groups – vehicle platooning, extended sensors, advanced driving, and remote driving [20]. These use cases have varying requirements, with some requiring direct vehicle-to-vehicle (V2V) communication while others take advantage of vehicle-to-infrastructure (V2I) communication. The following techniques will be specified in Rel-16 V2X work [21]:

- Sidelink design including signals, channels, bandwidth part, and resource pools. A sidelink is a direct link between NR devices without having to go through the base station.
- Resource allocation for sidelink. Two modes are supported as shown in Figure 3.14 – mode 1 is where the base station schedules the resource for

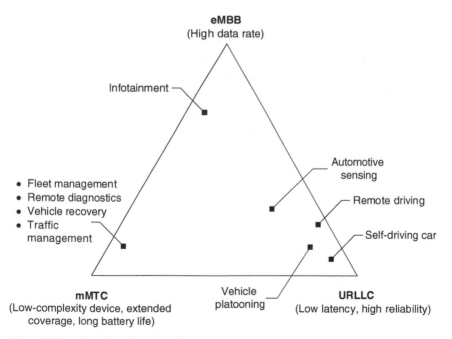

Figure 3.13 Examples of automotive use cases.

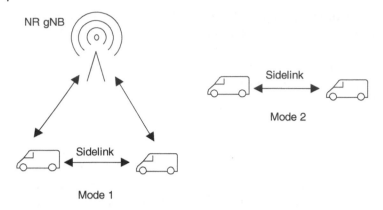

Figure 3.14 V2X resource allocation modes.

sidelink transmission and mode 2 is where the device autonomously selects the resource for sidelink transmission from a set of resources configured by the network.

- Sidelink synchronization mechanism including synchronization procedure when the devices are out of coverage of the base station.
- Sidelink physical layer procedures including HARQ, channel state acquisition and reporting, and power control.
- Sidelink L2/L3 protocols, signaling, and congestion control.

Vehicle platooning: Vehicle platooning is when a group of vehicles moves like a train with minimal distance between vehicles. This reduces fuel consumption, minimizes traffic, and allows a small number of drivers in case automated driving is enabled. To support platooning, vehicles must operate in a coordinated manner, sharing information such as speed, heading, acceleration, braking, etc. Communications are needed to allow vehicles to join or leave a platoon, exchange context-awareness messages to the group, and to inform nearby vehicles of the platoon. This information is shared using V2V communication without the need to go through the infrastructure.

Two types of platoons are being considered – normal and high density. In normal density platoons, the distance between vehicles is greater than 2 m. This requires end-to-end latency of approximately 25 ms to support vehicles traveling at 100 km/h. For high density platoons, the distance between vehicles is 1 m. This requires end-to-end latency of approximately 10 ms to support vehicles traveling at 100 km/h.

Based on the above use cases, the following requirements are being considered in 3GPP for V2X – able to support at least 30 broadcast messages per second, able to support at least 28 messages per second among a group of UEs, 10 ms end-to-end latency for message transfer among a group of UEs, capable of transferring messages among a group of UEs supporting V2X application with variable message payloads of 50–1200 bytes, support triggered and periodic transmission of small data packets, and support 90% reliability.

In addition to information sharing within the platoon using V2V, information sharing with the infrastructure using a Roadside Unit (RSU) is also beneficial. This can include sharing traffic information and road conditions, which may be accessed by other vehicles passing through the RSU at later times.

Automotive sensors and state map sharing: Sensor and state map sharing refers to the exchange of raw or processed data through local sensors or live video images among vehicles, RSUs, devices of pedestrian and V2X application servers. They can be used, e.g. for cooperative driving, intersection safety, emergency vehicle communication, and high-precision positioning. Requirements for this type of application include high date rate (approximately 25 Mbps per UE), low latency (less than 10 ms between V2X applications), high reliability, and high connection density for congested traffic (e.g. 3000 cars per square kilometer).

Another use case of sensor sharing is called Collective Perception of Environment where vehicles exchange real-time information within a local area. This can enhance knowledge about the environment and help to avoid accidents. For example, vehicles can periodically transmit information such as object classification, speed, and direction. Sensor data information can also be shared to inform vehicles of objects and obstacles that are not visible to the local sensors (e.g. objects around a curve).

Remote driving: Remote driving refers to a vehicle that is being remotely operated by a human. This can be complementary to autonomous driving and vehicle platooning. For instance, a truck can travel autonomously in a platoon on the highway but be operated remotely while in the city. This reduces the number of drivers required. Another example of remote driving is a bus that follows a predefined route where a human is needed to handle more complex scenarios such as people getting on and off the bus. For remote driving, on-board cameras can feed live video streams to the operator who will send commands to the vehicle.

To support remote driving, the following requirements are being considered – data rate up to 1 Mbps at downlink and 20 Mbps at uplink for speed of up to 250 km/h, and ultra-reliable transmission (at least 99.999%) at low latency (5 ms or less between the server and the vehicle).

Semi-automated or fully automated driving: Different levels of automated driving can be supported with different requirements. In 3GPP, the following use cases have been defined:

• Semi-automated driving at Society of Automotive Engineers (SAE) level 2 or 3 where coarse data exchange is sufficient. The required data rates are 0.55 Mbps for V2V and 0.5 Mbps for V2I. High reliability and connection density should be supported. The range may be as high as 0.6 km. Latency on the order of 100 ms is sufficient.

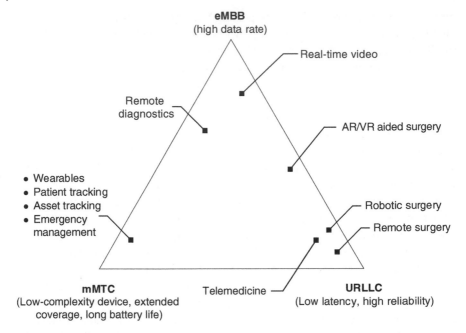

Figure 3.15 Examples of eHealth use cases.

- Full-automated driving at SAE level 4 or 5 where high-resolution data exchange (e.g. camera, LIDAR) is needed. The required data rates are 53 Mbps for V2V and 50 Mbps for V2I. High reliability and connection density should be supported. The range may be as high as 0.3 km. Latency on the order of 100 ms is sufficient.

In addition to supported automated driving, connected vehicles can also exchange information as part of collision avoidance. In this cooperative collision avoidance scheme, positioning information is exchanged among vehicles in addition to safety messages. They can be used for collision warning, intersection assistant, blind-spot assistant, and lane-change assistant. For this application, throughput of 10 Mbps, latency of less than 10 ms, and reliability of 99.99% are required between UE in proximity to perform coordinated driving maneuver at intersections.

3.3.3 eHealth

eHealth has emerged as an important vertical with 5G as an important enabler for various health-related areas. eHealth includes many use cases such as remote diagnostics, patient tracking, wearables, asset tracking, robotics, and telemedicine [22]. 5G NR provides the following advantages for deployment of eHealth services – high reliability and low latency, high data rates, the ability to support a massive number of devices, higher network efficiency, global connectivity under one platform, and the ability to offer differentiated services under one network. Examples of use cases from the eHealth vertical and how they are mapped to the 5G NR technologies are shown in Figure 3.15. These use cases are supported by Rel-15 NR features as described in this section.

Consider the challenging use case of remote surgery. This use case requires several components such as real-time video and audio streaming, monitoring of vital signs (e.g. temperature, blood pressure, heart rate, etc.), and haptic feedback [23]. Real-time video and audio streaming may require very high data rate (e.g. up 1 Gbps for 3D video) but is somewhat tolerant of the delay (e.g. latency up to 100 ms may be acceptable). Monitoring of vital signs generally requires low data rate and is also delay tolerant to a certain extent but might require a large number of sensors. Finally, haptic feedback requires very low latency (e.g. <10 ms) and very high reliability (e.g. 10e-4 packet error rate), but not very high data rate (e.g. 0.5 Mbps). These requirements cannot be satisfied with 4G LTE but they can be satisfied by 5G NR.

Another important eHealth use case is remote monitoring and diagnostics. This includes personal health wellness such as remote diagnostics and therapy for people with chronic diseases and vital sign monitoring using wearables. An example is a smart wearable glucose monitor where the data can be stored in the cloud and accessed by doctors. The wearable device provides eMTC or NB-IoT connectivity to the cloud without the need for a smartphone. Data collected can be used to improve wellness, monitor the effectiveness of treatments and suggest preventative measures. In this case, service requirements are similar to that for mMTC – support large number of devices, extended coverage, long battery life, and low-cost devices. Alternatively, using indoor RF sensing technology one can monitor presence, heart rate, activity pattern and location for monitoring elderly patients. These could be implemented using 5G features such as centimeter level localization, URLLC and massive MTC technologies.

Some key eHealth requirements and relevant 5G NR enablers are:

- Support for mMTC, including the ability to support a large number of devices and extended coverage. The network must be highly scalable and cost effective to support large device density within an area. In 5G, NB-IoT and eMTC technologies offer the ability to support large number of devices. Additional capacity can be easily added by deploying additional carriers or by using larger bandwidth. Since other services such as eMBB and URLLC can coexist with mMTC in a seamless manner, additional capacity can be added or removed via dynamic resource allocation. This allows for a very efficient to support mMTC traffic within the 5G system. In addition, NB-IoT and eMTC support coverage enhancement of up to

20 dB compared with 4G LTE. This allows 5G mMTC services to be deployed in previously unavailable locations such as deep within a building or inside elevators.

- Energy efficiency to support battery life of 10–15 years for wearables or implant devices. 5G provides a very lean communication protocol and physical layer structure such that devices can be maintained in low-power mode until needed. Devices can wake up and quickly transmit data before going back to sleep again. In addition, NB-IoT and eMTC technologies support the use of a wake-up signal where it is possible for devices to remain in low-power mode until woken up by the network for data transmission.
- Latency reduction to support new use cases such as robotics, remote surgery, remote control, and telemedicine. For example, remote surgery may require end-to-end latency on the order of 10–20 ms, which can be delivered by 5G URLLC. Dedicated radio and network resources can be provisioned to support this service via network slicing.
- Reliability improvement to support critical use cases such as remote surgery and remote diagnostics. Reliability refers not just to the quality of the link but also the ability to react to failures, such as continuous service is always ensured to the end user. 5G NR can deliver very high reliability (e.g. 10e-5 packet error rate). In addition, it can support 0 ms mobility interruption time, which allows uninterrupted service continuity.
- Network flexibility including the ability to differentiate and provide different QoS to different services running in the same network. Network slicing can be configured to satisfy the requirements for various services even within the same use case (e.g. remote surgery may require different QoS for haptic, monitoring, and multimedia components).

3.4 Conclusion

In this chapter, we provided an overview of NR radio interface design and how it can be tailored to meet the requirements of different verticals using three key NR components: eMBB, mMTC, and URLLC. Three specific vertical industries, namely IIoT, V2X, and eHealth, were discussed. It was shown how these verticals could benefit from various features of 5G NR technology.

Acknowledgment

The authors would like to thank Eugene Visotsky for providing the performance results included in the chapter.

Acronyms

BLER	Block Error Rate
CQI	Channel Quality Indicator
CRI	CSI-RS Resource Indicator
CSI	Channel State Information
CSI-RS	Channel State Information Reference Signals
DCI	Downlink control information
DMRS	Demodulation Reference Signal
eDRX	Extended discontinuous reception
ECID	Enhanced Cell ID
eMBB	Enhanced mobile broadband
eMTC	Enhanced machine type communication
eNB	Enhanced Node B (base station)
FDD	Frequency Division Duplex
gNB	Next generation Node B (base station)
IoT	Internet of Things
LDPC	Low-density parity-check
LTE	Long-Term Evolution
MCL	Maximum Coupling Loss
MCS	Modulation and Coding Scheme
mMTC	Massive machine type communication
MU-MIMO	Multi-user multiple-input multiple-output
NB-IoT	Narrowband Internet of Things
NR	New Radio
OTDOA	Observed Time Difference of Arrival
PBCH	Physical broadcast channel
PDCCH	Physical downlink control channel
PDSCH	Physical downlink shared channel
PMI	Precoding Matrix Indicator
PRACH	Physical random access channel
PRB	Physical resource block
PSS	Primary Synchronization Signal
PTRS	Phase Tracking Reference Signal
PUSCH	Physical uplink shared channel
RI	Rank Indicator
RRC	Radio resource connection
RSU	Roadside Unit
SAE	Society of Automotive Engineers
SC-PTM	Single-cell point-to-multipoint
SCS	Subcarrier spacing

SRB	Signaling Radio Bearer
SRS	Sounding Reference Signal
SSB	Synchronization Signal Block
SSS	Secondary Synchronization Signal
SU-MIMO	Single-user multiple-input multiple-output
TDD	Time Division Duplex
TSC	Time-sensitive communication
UCI	Uplink control information
UE	User equipment
URLLC	Ultra-reliable low-latency communications
V2I	Vehicle-to-infrastructure
V2P	Vehicle-to-pedestrian
V2V	Vehicle-to-vehicle
V2X	Vehicle-to-everything
VoLTE	Voice over Long-Term Evolution
ZF	Zero Forcing

References

1 TR 38.912 (2017). Study on New Radio (NR) Access Technology, V14.0.0, March 2017.

2 TR 38.913 (2017). Study on Scenarios and Requirements for Next Generation Access Technologies, V14.3.0, June 2017.

3 Ghosh, A. (2018). 5G New Radio (NR): Physical Layer Overview and Performance. IEEE Communication Theory Workshop (May 2018).

4 Ghosh, A. (Sep 2016). 5G mmWave revolution. *Microwave Journal* 59 (9): 22–36.

5 RP-190712 (2019). Revised WID: Integrated Access and Backhaul for NR, 3GPP RAN#83, Shenzhen, China, March 2019.

6 Chen, J., Chen, Y., Jayasinghe, K. et al. (2017). Distributing CRC Bits to Aid Polar Decoding. IEEE Globecom Workshops, Singapore, pp. 1–6.

7 TR 38.840 (2019). Study on UE Power Saving in NR, V1.0.0, March 2019.

8 Li, Z., Uusitalo, M.A., Shariatmadari, H., and Singh, B. (2018). 5G URLLC: design challenges and system concepts. 15th International Symposium on Wireless Communication Systems, Lisbon, pp. 1–6.

9 Ji, H., Park, S., Yeo, J. et al. (2018). Ultra-reliable and low-latency communications in 5G downlink: physical layer aspects. *IEEE Wireless Communications* 25 (3): 124–130.

10 TR 36.888 (2013). Study on Provision of Low-Cost Machine-Type Communications (MTC) User Equipments (UEs) Based on LTE, V12.0.0, June 2013.

11 TR 45.820 (2015). Cellular System Support for Ultra Low Complexity and Low Throughput Internet of Things, V2.1.0, August 2015.

12 Ratasuk, R., Mangalvedhe, N., and Ghosh, A. (2015). Overview of LTE enhancements for cellular IoT. IEEE PIMRC 2015, Hong Kong.

13 Rico-Alvarino, A., Vajapeyam, M., Xu, H. et al. (2016). An overview of 3GPP enhancements on machine to machine communications. *IEEE Commununications Magazine* 24 (6): 14–21.

14 Ratasuk, R., Mangalvedhe, N., Bhatoolaul, D., and Ghosh, A. (2017). LTE-M evolution towards 5G massive MTC. IEEE Globecom Workshops, Singapore, pp. 1–6.

15 R1-1904170 (2019). Coexistence of eMTC with NR, RAN1#96bis, Xi'an, China, April 2019.

16 TR 22.261 (2018). Service Requirements for the 5G System, V16.5.0, September 2018.

17 RP-190728 (2019). New WID: Support of NR Industrial Internet of Things (IoT), 3GPP RAN#83, Shenzhen, China, March 2019.

18 RP-190726 (2019). New WID: Physical Layer Enhancements for NR Ultra-Reliable and Low Latency Communication (URLLC), 3GPP RAN#83, Shenzhen, China, March 2019.

19 TR 38.855 (2019). Study on NR Positioning Support, V16.0.0, March 2019.

20 TR 22.886 (2018). Study on Enhancement of 3GPP Support for 5G V2X Services, V16.1.1, September 2018.

21 RP-190766 (2019). New WID on 5G V2X with NR Sidelink, 3GPP RAN#83, Shenzhen, China, March 2019.

22 Politis, C., Thuemmler, C., Grigoriadis, N. et al. (eds.) (2016). A New Generation of e-Health Systems Powered by 5G. Wireless World Research Forum (November 2016).

23 Soldani, D. Fadini, F., Rasanen, H. et al. (2017). 5G mobile systems for healthcare. IEEE 85th Vehicular Technology Conference, Sydney, pp. 1–5.

4

Effects of Dynamic Blockage in Multi-Connectivity Millimeter-Wave Radio Access

Vitaly Petrov, Margarita Gapeyenko, Dmitri Moltchanov, Andrey Samuylov, Sergey Andreev, and Yevgeni Koucheryavy

Tampere University, Tampere, Finland

Abstract

This chapter outlines the methods to capture and characterize the blockage effects in cellular millimeter-wave (mmWave) systems together with the state-of-the-art techniques to mitigate the negative effects of dynamic blockage on both user- and network-centric performance indicators. It first discusses the key aspects of mmWave signal blockage by various objects in urban deployments and its consecutive impact on link-level performance. Further, the chapter summarizes mathematical methods to model the dynamic blockage processes in different Internet-of-Things (IoT) scenarios and deployment configurations, such as massive augmented and virtual reality setups, connected vehicles, etc. The presented results can be used in design and optimization of cellular mmWave systems for the emerging delay-, availability-, and reliability-sensitive services across many 5G verticals. The chapter also addresses possible solutions to mitigate blockage in large-scale 5G-grade consumer IoT deployments equipped with mmWave New Radio connectivity.

Keywords *5G verticals; cellular millimeter-wave systems; dynamic blockage avoidance; Internet-of-Things scenarios; multi-connectivity millimeter-wave radio access; network-centric performance indicators; user-centric performance indicators;*

4.1 Introduction

Recent adoption of millimeter-wave (mmWave) radio access is becoming instrumental to support high-rate connectivity for 5G verticals. This effort is spearheaded by 3GPP as part of their New Radio (NR) initiative [1]. The utilization of

5G Verticals: Customizing Applications, Technologies and Deployment Techniques, First Edition.
Edited by Rath Vannithamby and Anthony C.K. Soong.
© 2020 John Wiley & Sons Ltd. Published 2020 by John Wiley & Sons Ltd.

mmWave radio in 5G and beyond systems is expected to provide substantial improvements in both peak and average user capacity, energy efficiency, and latency, among other parameters. Despite many decisive benefits, the adoption of mmWave communication requires a major upgrade of multiple communication protocols due to the need to operate with large-scale antenna arrays, inherent directionality of data transmissions at these frequencies, and order-of-magnitude larger bandwidth.

In addition, mmWave cellular systems face certain challenges related to providing uninterrupted connectivity in dynamic environments. This is because the quality of fragile mmWave connections depends heavily on the presence of a line-of-sight (LoS) path between the communicating entities [2]. The latter is susceptible to blockage by various obstacles, such as buildings, vehicles, and human bodies [3]. The understanding of blockage effects is particularly crucial for emerging mission-aware applications that involve moving connected objects including vehicles and drones [4].

In this chapter, we outline the methods to capture and characterize the blockage effects in cellular mmWave systems together with the state-of-the-art techniques to mitigate the negative effects of dynamic blockage on both user- and network-centric performance indicators. First, in Section 4.2, the key aspects of mmWave signal blockage by various objects in urban deployments and its consecutive impact on link-level performance are discussed. Further, mathematical methods to model the dynamic blockage processes in different Internet-of-Things (IoT) scenarios and deployment configurations, such as massive augmented and virtual reality (VR) set-ups, connected vehicles, etc., are summarized in Section 4.3. The presented results can be used in design and optimization of cellular mmWave systems for the emerging delay-, availability-, and reliability-sensitive services across many 5G verticals.

The second part of this chapter – Sections 4.4–4.6 – addresses possible solutions to mitigate blockage in large-scale 5G-grade consumer IoT deployments equipped with mmWave NR connectivity. Particularly, the ways to exploit multi-connectivity (MC) operation – a recent 3GPP technique to enable simultaneous user interaction with multiple radio access nodes – for improved session continuity under dynamic blockage are discussed in Section 4.4. The concept of bandwidth reservation to tolerate abrupt fluctuations in the network load caused by dynamic blockage is introduced and explained in Section 4.5. Finally, proactive link selection in MC mmWave access to predict the imminent link blockage and minimize its harmful effects is summarized in Section 4.6. Concluding remarks on dynamic blockage modeling and mitigation for various 5G verticals are offered in Section 4.7.

4.2 Blockage Effects in 5G Millimeter-Wave Cellular Communication

4.2.1 Millimeter-Wave Link Blockage at a Glance

There are several types of blockage in mmWave-based IoT deployments. The first one is caused by large stationary objects (e.g. buildings) [5]. This type of blockage

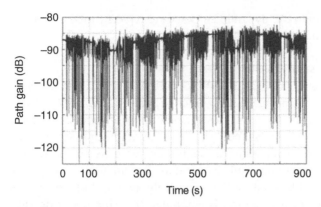

Figure 4.1 Dynamics of mmWave signal attenuation due to blockage. *Source:* [7].

was also considered in the previous generation of mobile systems. Another type of blockage, which was not considered before, is caused by small-scale objects including so-called self-blockage [5]. The latter blockage may be dynamic (humans, cars, etc.) or static (lampposts, trees, etc.). Self-blockage occurs due to occlusions caused by the user itself (e.g. the hand is covering the antenna). This problem may be resolved by utilizing smarter antenna configurations, which decrease the chances of blocking all of the antenna elements [6].

A number of experiments demonstrated that blockage by the human body and cars may introduce an additional 20–40 dB loss at e.g. 28 GHz as shown in Figure 4.1 [7]. In [2], the experiments conducted in an open pedestrian street confirmed the attenuation due to human-body blockage of up to 25 dB, having a fading duration of up to 200 ms.

4.2.2 Blockage Modeling Methodology

In order to consider link blockage in future studies, it is essential to be able to characterize it by simulation and/or analysis. Below, two kinds of typical dynamic blockage situations in urban IoT scenarios (by a human body and a car) are considered. However, the sources of blockage are not limited to them, as it might be caused by any other object with a sufficiently large size for mmWave frequencies. Further, a blockage modeling methodology is presented.

4.2.2.1 Geometric Representation of Blocking Objects
First, it is important to represent a human body or a car in a tractable way. In recent works, the human body is typically modeled as a cylinder with a certain height and width [8]. Moreover, some modeling tools such as ray launching require the knowledge of the physical properties of the objects. The same study demonstrated that a human body may be approximated by a cylinder filled with

water. Cars as another source of dynamic blockage are typically modeled as rectangular shaped cubes, not transparent to mmWave signal [9].

4.2.2.2 Attenuation Caused by Blocking Objects

In addition to experimental studies, 3GPP proposed another way to consider attenuation caused by humans and cars as described in ([5]; section 7.6.4) and named Model *A* and Model *B*. Model *A* is a stochastic approach where the blockers are represented as rectangular shapes. The positions of multiple screens that fully block the mmWave signal passing through them are generated randomly by using well-known distributions and the resultant attenuation is computed. In Model *B*, the screens with certain width and height are physically placed on the map, and after that the attenuation is computed by using the knife edge diffraction model.

4.2.2.3 Channel Models

The first one, the so-called LoS-based model, delivers the path loss in the case of LoS and non-LoS (nLoS) subject to a chosen probability. This model provides the average received power at the receiver (Rx) side without specifying the direction of the reflected component ([5]; table 7.4.1-1). The second model is an advanced algorithm developed specifically to consider the multipath propagation at mmWave frequencies ([5]; section 7.5). In this model, individual clusters are characterized by their own angles of arrival and departure, and each of them has its own power and delay.

4.2.2.4 Blockage States

The following states are typically differentiated between: *LoS/nLoS* (LoS path occlusion by large-scale objects, such as buildings) and *blocked/non-blocked* (LoS path occlusion by small-scale objects, such as vehicles and pedestrians). In stochastic models, a transition from the LoS to nLoS state is performed with the probability of LoS from ([5]; table 7.4.2). The choice of the channel model is of special importance to determine the received power after the blockage has occurred. For a first-order analysis, a simple LoS-based propagation model suffices if supplemented with additional attenuation due to blockage [6]. For a more precise evaluation, every state is given its own path loss that is further used to establish the received power at the device side. A deeper approach is the use of a multipath channel model, where in case of blockage of one link the device might choose the second strongest non-blocked cluster.

4.2.3 Accounting for mmWave Blockage

In order to incorporate blockage into the system-level performance evaluation, one may choose to simulate it directly by physically mapping the objects with their properties as discussed earlier. However, as it was shown in [10], this approach increases the simulation time significantly. Another method for modeling blockage is the use of analytical tools. In this case, blockage is represented by an appropriate analytical process and the desired metrics of interest are derived. Finally, a mixture of analysis and simulation is the third possible option. In more detail, instead of modeling the blockage directly, one may use the distribution of time spent in blockage and the probability of blockage by applying them for faster simulation.

4.2.4 Summary

This section has shown that mmWave signal blockage is an important consideration for future mmWave systems, which requires focused attention. Therefore, capabilities to model the dynamics of the blockage process in simulation-based and analytical frameworks are essential. These are discussed in more detail in the following section.

4.3 Modeling Consumer 5G-IoT Systems with Dynamic Blockage

In this section, the details of blockage modeling in various consumer 5G-grade IoT scenarios are offered. These differ in the underlying assumptions regarding the blocker mobility (static or dynamic) and considerations on the spatial and temporal correlation of blockage.

4.3.1 Spontaneous Public Event

The first scenario corresponds to a large-scale public event attracting pedestrians with high-end wearable or handheld devices, e.g. augmented reality (AR) and VR glasses [11]. The scenario features neighboring pedestrians acting as static blockers, displaying no mobility of the target IoT device and weak correlation in the blockage process.

In this setup, one access point (AP) and one IoT device are separated by distance r_0 and located at height h_T and h_R. The size of Rx is assumed to be infinitesimally small. Static blockers distributed around the user equipment (UE) form a Poisson point process (PPP) with density λ_B. In this case, blockers are assumed to be human bodies modeled as cylinders with height h_B and base diameter d_B. The problem to calculate the blockage probability is discussed below and its detailed explanation is given in [12].

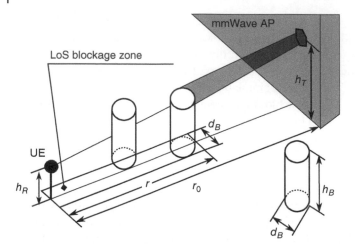

Figure 4.2 LoS blockage zone.

The term *LoS blockage zone* is defined as illustrated in Figure 4.2. It is the zone of a rectangular shape where the presence of a blocker will cause a link occlusion, thus leading to blockage. The width of this zone is equal to the diameter of blocker d_B, while the length is equal to r. As the effective aperture of a mmWave antenna system is relatively small, the actual signal propagation can be modeled as a segment between the nodes. Hence, the blockage zone has a rectangular shape and does not follow the trapezoid shape of the mmWave beam.

By recalling the property of PPP regarding the number of points in the area of interest, one may derive the probability of non-blockage $p_{nB}(x)$ by calculating the probability of having no blockers in the LoS blockage zone as [12]

$$p_{nB}(x) = e^{-d_B \lambda_B \left[x \frac{h_B - h_R}{h_T - h_R} + d_B/2 \right]}. \tag{4.1}$$

Therefore, the probability of blockage is $p_B(x) = 1 - p_{nB}(x)$. As a result of these derivations, one may find the probability of blockage as a function of various system parameters. Figure 4.3 illustrates the probability of blockage depending on the height of the AP and the distance between the AP and the IoT UE. The resulting probability of blockage increases as the distance between the nodes grows. However, one may notice that by increasing the height of the AP the probability of blockage decreases.

4.3.2 Moving Through the Crowd

The second scenario reflects the case where a UE moves through the crowd of blockers. It is featured by a mobile IoT device and static blockers. In this scenario, spatial correlation between two different positions of the IoT device is modeled to

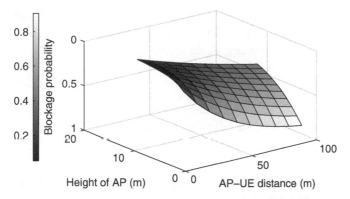

Figure 4.3 Blockage probability as a function of AP height and AP–UE distance.

Figure 4.4 2D view of UE position.

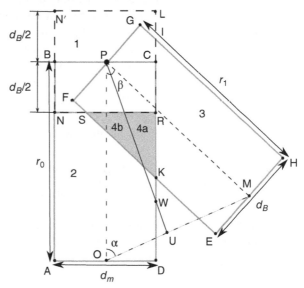

establish the conditional probability of blockage. The derivation method is based on a combination of the probability theory and geometric considerations; it is described in detail in [13]. A geometric representation of this scenario is offered in Figure 4.4. The AP is located at point P, the first location point of the IoT device is O, the second point is M. The two points O and M can be separated in time, as one IoT device moves from O to M, or in space, when two neighboring receivers are located at points O and M. The problem is to find the conditional probability, p_{ij}, of the IoT device being in state i at point O and in state j at point M (0 and 1 correspond to the non-blocked and blocked state, respectively).

There are two possible events that may occur, non-blockage and blockage at point M conditioned on the state at point O, which organize the sample space. Therefore, the sum of their probabilities is equal to one and

$$p_{00} = 1 - p_{01}, \qquad p_{10} = 1 - p_{11}. \tag{4.2}$$

When calculating the joint probability of the following non-blocked state at M and blocked/non-blocked state at O, the following holds

$$p_{nB,M} = p_{00}p_{nB,O} + p_{10}p_{B,O}, \tag{4.3}$$

where $p_{nB,M}$, $p_{nB,O}$, and $p_{B,O}$ are derived using Eq (4.1). Therefore, in order to produce p_{ij}, it is sufficient to evaluate p_{00}.

As already mentioned, the method of deriving the conditional probability is based on geometry as illustrated in Figure 4.4, so that the two LoS blockage zones for the points O and M are subdivided into smaller zones. The width of the LoS blockage zone is equal to the diameter of a blocker d_m, while its length equals the distance between P and O/M. There are four major zones as depicted in Figure 4.4. Zone 1 was excluded from the analysis due to its small dimensions and complex consideration. Zones 2 and 3 are the ones impacting the LoS at points O and M, respectively. The most important zones named 4a and 4b are the ones affecting both links simultaneously. Every zone is analyzed separately to establish the probability of it having blockers. Finally, the conditional probability p_{00} is derived as follows

$$p_{00} = \mathbf{P}\big[\text{nB at } M \mid \text{nB at } O\big] = \frac{\mathbf{P}\big[\text{nB at } M \cap \text{nB at } O\big]}{\mathbf{P}\big[\text{nB at } O\big]}. \tag{4.4}$$

In addition to the conditional probability, the proposed methodology allows to assess the correlation distance. Figure 4.5 shows the correlation distance as a function of angle α, which demonstrates the position of M with respect to O. It is observed that by increasing the angle α from 0 to $\pi/2$ the correlation distance decreases. This is explained by the fact that the common zone between the two LoS blockage zones becomes smaller. Using this methodology and varying the angle α, one may identify the shape of the zone where the correlation between two points exists. Outside of this zone the probability of being blocked/non-blocked at point M does not depend on the blocked/non-blocked state at point O.

4.3.3 AR Sessions in Dense Moving Crowd

The third scenario captures the use case where a static UE is surrounded by a moving pedestrian crowd. For instance, this may be a sightseeing user with AR glasses or high-resolution camera who simultaneously streams high-rate video.

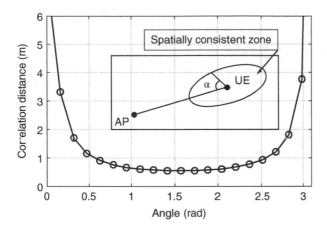

Figure 4.5 Correlation distance as a function of angle α.

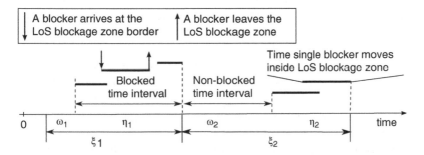

Figure 4.6 Renewal process associated with LoS blockage.

This scenario features a static UE surrounded by a dynamic field of blockers. The blockage process itself experiences temporal correlation. The focus in analyzing this scenario is the blockage state evolution in the time domain. In this scenario, the IoT device is assumed to remain static while the blockers are moving around. The mobility of blockers may follow different analytically tractable models. In this particular example, random direction mobility (RDM) is assumed.

To establish the conditional probability of blockage at time t_1 given that the IoT device was in blocked/non-blocked condition at time t_0, we employ the renewal theory [10]. The time evolution of the blocker dynamics as they enter and leave the LoS blockage zone is depicted in Figure 4.6. The process of entering the LoS blockage zone is Poisson in time. The time that one blocker spends in the LoS blockage zone follows the distribution derived in [10]. By analogy with the queuing theory, we notice that the process of entering and leaving the blockage zone is similar to the process of packet arrivals.

The methodology to derive the conditional probability of blockage is based on the time domain convolutions. The blockage process is represented by a number of events that interchange one another and the following is true:

$$p_{00}(\Delta t) = \sum_{i=0}^{\infty} \mathbb{P}\{A_i(\Delta t)\}, \qquad p_{01}(\Delta t) = \sum_{i=1}^{\infty} \mathbb{P}\{B_i(\Delta t)\}, \tag{4.5}$$

where $A_i(t)$ are the events reflecting the case of starting in a non-blocked interval at t_0 and ending in a non-blocked interval after some $\Delta t = t_1 - t_0$, while having exactly i ($i = 0, 1, ...$) blocked periods during Δt. Similar reasoning is applied to $B_i(t)$.

4.3.4 Connected Vehicles

The last setup addressed in this section is the use of high-end IoT systems in a vehicular environment. One of the typical mission-critical use cases addressed in detail in [14] is described below. An ambulance vehicle running a critical application, e.g. remote patient assessment, moves along a designated lane at constant speed. The 5G NR APs are assumed to be deployed along the street on the building sides to support extreme traffic volumes with stringent delay and data rate requirements [15]. In this environment, vehicles moving along other lanes become blockers for the signal path between the IoT device and its associated NR AP. The authors in [14] represent blockers as screens having specific height and length that coincide with the dimensions of typical cars.

A blockage period is fully determined by the distribution of the vehicle length, whose height exceeds the LoS height. Assuming the independence of vehicle heights on a lane, the probability density function (pdf) of the non-blocked distance is given by

$$f_{d_{t,i}}(x) = q_0 f_{d_t}(x) + \sum_{i=1}^{\infty} q_i \left[f_{l_B}(x) * f_{d_t}(x) \right]^{(i)}, \tag{4.6}$$

where the superscript (i) denotes the i-fold convolution, and $f_{l_B}(x)$ and $f_{d_t}(x)$ are the pdfs of the vehicle length and inter-vehicle distance, respectively. Here, $q_i = 1/\lambda_{E,i}$ ($i = 0, 1, 2$), where $\lambda_{E,i} = p_{B,i}\lambda_i$ is the effective intensity of vehicles on lane i that block the LoS path. The mean non-blocked distance is readily obtained by $(E[l_b] + E[d_i])/q_i$. The pdf of the non-blocked time can be obtained by scaling the resulting distribution with the relative speed of the tagged UE and blocking vehicles.

In [14], an analogy with the queuing theoretic model has been provided by identifying an approximation with the Laplace transform that obeys the Kendall functional equation. Hence, the approximation of the mean is readily given by [16] as

$$E[T_B] = \frac{E[T_B^*]E[T_L]}{E[T_L] - E[T_B^*]}. \tag{4.7}$$

4.3.5 Summary

The choice of a particular model to account for dynamic blockage heavily depends on the deployment configuration, node mobility, and scenario specifics. As can be seen in Figure 4.3, dynamic blockage has a strong impact on mmWave connectivity. Therefore, there are techniques to combat the negative effects of dynamic blockage in consumer IoT scenarios. These techniques – MC, bandwidth reservation, and proactive link selection – are reviewed in the following three sections. The models outlined in this section can be considered as building blocks for more realistic scenarios. For example, a natural extension is to address a combination of models from Sections 4.3.2 and 4.3.3 where the UE moves through a dense mobile crowd.

4.4 Dynamic Multi-Connectivity

4.4.1 Multi-Connectivity at a Glance

MC is one of the recent solutions introduced for 5G NR systems in order to improve the session rate and reliability, especially in dense urban environments [17]. This technique stems from the dual connectivity proposed for the previous generations of cellular systems and allows the UE to be connected to multiple APs simultaneously [18]. According to the studies in [19–21], the utilization of MC is particularly beneficial for mitigating the dynamic blockage effects in large-scale IoT systems operating over the mmWave bands. In such deployments, sudden connection interruptions do not immediately lead to the UE outage, since the data session can be almost instantly rerouted to one of the other momentarily non-blocked APs (Figure 4.7).

While MC can bring decisive benefits to 5G NR in terms of session continuity (and thus effective data rate and latency), its application raises a number of concerns related to efficient integration of MC into the 5G systems, in particular: "How to combine multiple data threads coming from/to UE simultaneously via different APs?" "How many simultaneous connections are appropriate in different IoT deployments?" "How to properly account for MC when evaluating the system-level performance?" "How to select and prioritize the APs out of those currently available?" and many more.

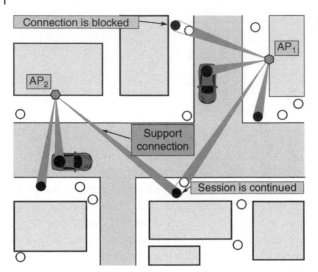

Figure 4.7 Use of mmWave MC in urban scenarios.

Today, the standardization of MC for 5G NR is still ongoing [17] where many important engineering choices require further clarification. However, certain facts about the utilization of MC in typical consumer IoT deployments have already been established in recent research reports.

4.4.2 Optimizing the Degree of Multi-Connectivity

One of the key engineering considerations related to the MC integration into 5G is the choice of the maximum number of simultaneous links to be handled by the UE – so-called *degree of MC*. Here, the principal trade-off arises where the higher degree contributes to improved reliability, whereas leading to increased computation and energy overheads to establish and maintain connections with multiple APs. In addition, higher degrees of MC also challenge the channel utilization, as an increased fraction of radio resources is allocated to beam tracking and service signaling with the APs.

The authors in [22] addressed this issue by studying the impact that the degree of MC has on the high-rate IoT session continuity in the typical 3GPP urban scenarios. They modeled the NR system at 28 GHz with randomly deployed APs and UEs as well as certain densities of the human crowd dynamically blocking the mmWave signal. It was assumed that in case of a link blockage to one AP, the UE operates with another one having the best signal quality. Therefore, the MC is assumed to provide improved spatial diversity. In that work, the authors assumed ideal synchronization. This is an important matter and interested readers may refer to [17]. It was therefore shown that introducing a single additional link (dual connectivity) at the cell edge reduces the outage probability by up to 72% (Figure 4.8).

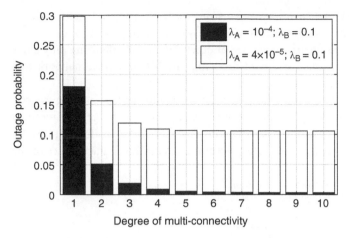

Figure 4.8 Impact of the degree of MC on outage probability.

A further increase in the degree of MC leads to much smaller gains. After connecting to approximately four mmWave APs, the UE ceases receiving any notable benefits from increased degree of MC. Therefore, the recommended degree of MC for the consumer IoT urban deployments is in the range from 2 to 4 depending on the network configuration and UE constraints in terms of performance, battery lifetime, and beam tracking capabilities.

4.4.3 Modeling 5G NR Systems with Multi-Connectivity

The introduction of MC imposes additional challenges in terms of modeling the 5G system behavior and performance. As each of the MC-capable UEs is able to be connected to many APs simultaneously, the overall state space expands rapidly with the degree of MC. Consequently, the complexity of performance evaluation increases and calls for certain enhancements in the conventional tools used to model 5G-grade mmWave systems.

Particularly, simulation-based tools (both time- and event-driven) have to capture the event of a data session rerouting and splitting across several APs. Therefore, UE-centric metrics (such as data rate, drop rate, latency, etc.) need to be aggregated across multiple links with several APs. In addition, the queues used to store the packets waiting for a transmission opportunity should be implemented in a way that supports on-the-fly data rerouting to another logical and physical channel. All the above notably increases the complexity and performance demands of the system-level simulators capable of modeling the advanced MC techniques.

In order to improve the scalability and reduce the computation time, it is also possible to model MC-enabled 5G systems analytically. Here, the complexity of the associated mathematical framework also grows as extra dependencies between the system states appear: the details of the session service process at one of the APs influence the session arrival process at other APs. Today, there are only a few mathematical frameworks capable of modeling the MC mmWave network to a certain extent [20, 21, 23].

One of the possible approaches is to design a sophisticated queuing system, where the sessions interrupted during their service at one of the APs are rerouted into the arrival flow of other APs, by following the selected MC policy (Figure 4.9). It was shown in [20] that under certain assumptions on the characteristics of the

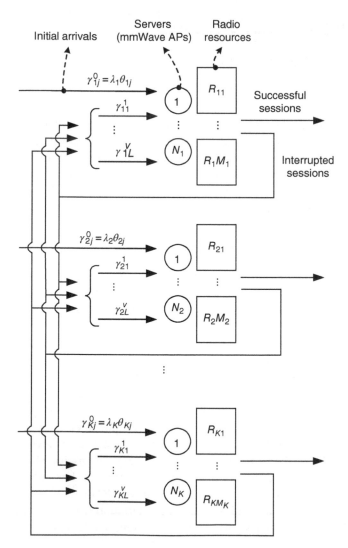

Figure 4.9 Queuing network framework for MC-capable systems.

session arrival flow, it is possible to derive the key performance indicators with a semi-analytical iterative approach. Depending on the target IoT application and scenario, the approach presented in [20] can be adjusted to support multiple session types, heterogeneous radio resources, and different deployment configurations. Meanwhile, the complexity of the developed solution is much higher than that of the frameworks modeling 5G systems without MC.

4.4.4 Impact of Multi-Connectivity Policy

While dual connectivity already brings notable performance improvements to the 5G NR systems, an additional question arises when the degree of MC becomes greater than two. It is particularly related to the criteria used to prioritize the mmWave APs and manage the data transmissions among them subject to possible dynamic blockage. Here, the system-level aspects of the 5G deployment are of particular interest, since some of the MC policies may result in many UEs temporarily selecting a particular AP to handle their traffic, thus overloading this AP and causing service degradation. A further difficulty is that the signal quality degradation may occur unexpectedly and require immediate reaction by the UE due to stringent latency constraints of the 5G-grade IoT traffic.

It was recently demonstrated that the choice of the MC policy has a significant impact on both user- and network-centric performance indicators. Particularly, an envisioned massive IoT scenario with the 5G NR network handling hundreds of AR sessions from mobile UEs in a large city square was studied in [20]. The authors selected the Times Square in New York City, NY, USA as a representative deployment and applied a combination of analytical methods as well as ray-based, link- and system-level computer simulations to compare several MC policies:

1) *No reconnection (Baseline, NORECON).* The baseline strategy where the AR UE is always connected to a single mmWave AP.
2) *Dual connectivity (DCON).* Each AR UE is simultaneously connected to two mmWave APs. The UE reacts to the dynamic blockage with one AP by rerouting its data session to another AP.
3) *MC, blockage (MCON, bkg).* This is an extension of the previous policy where the degree of MC is set to three. The AR UE selects an alternative AP in a randomized manner in order to avoid possible overloads of its nearest mmWave AP.
4) *MC, throughput (MCON, tpt).* A modification of the previous policy where the estimated maximum throughput is used as the criterion to rank the APs (as opposed to the probability of link blockage).
5) *MC, both blockage and throughput (MCON, bkg + tpt).* An evolved version of the preceding two strategies. The estimated *effective* throughput (maximum throughput weighed with the probability of link blockage) is used as the AP ranking criterion.

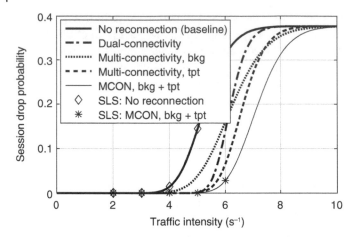

Figure 4.10 Effects of MC policy on session drop probability.

For the given MC policies, Figure 4.10 reports on the continuous session drop probability as a function of the AR traffic arrival intensity for a moderately dense crowd with a pedestrian positioned at every $2\,m^2$, on average. The figure presents the metrics estimated analytically and the data produced with the system-level simulator (SLS). As can be seen, the continuous session drop probability does not grow beyond a certain limit. The respective threshold heavily depends on the density of humans as well as the system parameters ≈ 0.38 for the considered setup. The reason behind this behavior is that as the traffic intensity grows, the considered system first refrains from accepting new sessions and only drops the already accepted ones if they request prohibitively high resources due to blockage.

In this figure, it is clear that the difference in the outage probability and maximum tolerable session arrival rate (indirectly characterizing the aggregate system throughput) between the MC policies can easily reach 100%. More advanced policies, such as *MCON, bkg + tpt*, notably outperform their simpler counterparts. In addition, depending on the current network loading, one or another MC policy may offer an advantage. Therefore, the ability to dynamically adjust not only the set of APs used for MC but the MC policy itself has certain benefits in practical IoT deployments.

4.4.5 Summary

MC is a novel technique in 5G systems, which is currently considered as one of the most promising solutions to mitigate dynamic blockage of mmWave links and hence improve the reliability and data rates in highly loaded scenarios. In this section, several aspects of MC modeling and deployment in 5G mmWave networks were discussed and the underlying trade-offs related to selecting the degree of MC as well as the appropriate MC policy were reviewed. While the process of MC development and standardization as part of 5G NR remains active, these preliminary performance predictions confirm its high potential to non-incrementally improve both user- and network-centric performance, thus calling for follow-up studies on the subject.

4.5 Bandwidth Reservation

4.5.1 Session Continuity Mechanisms

In mmWave 5G systems, frequent link state changes from non-blocked to blocked result in abrupt fluctuations of bandwidth requirements during the session service time. This may lead to a drop of an ongoing session having certain data rate requirements, since its serving mmWave AP may not have sufficient radio resources to maintain it. From the quality of user experience perspective, it may be preferable to reject a session at the moment of its arrival rather than drop it during service [24]. Hence, in the presence of uncertainty about the channel state at each active UE, it is reasonable to reserve a fraction of system resources for the ongoing sessions by only making a fraction of total resources available for the new session arrivals. Once a session is accepted, the entire set of resources is made accessible to it. This concept may improve session continuity at the expense of increased drop probability for new sessions.

4.5.2 Concept of Bandwidth Reservation

With the aim of improving session continuity the authors in [25] introduced the concept of bandwidth reservation for mmWave NR systems. They considered a standalone NR AP with a set of UEs distributed randomly across its service area. Farther away UE locations are assumed to be limited by the distance where a UE experiencing blockage still resides in non-outage conditions. The blockage is assumed to be dynamic, that is, humans move over the area of interest according to a stationary mobility model. The durations of blocked and non-blocked intervals follow from [10]. The traffic is assumed to be non-elastic: the data rate requested by an application at the session initialization phase needs to be maintained throughout the entire active session.

In the described system, bandwidth reservation is made available by ensuring that a fraction γB, $\gamma << 1$, of all the bandwidth available at the NR AP is not accessible for new session arrivals. However, the entire bandwidth B can be used to serve sessions that have already been accepted by the system. This procedure implicitly prioritizes sessions, which were selected for service over any new arrivals, thus making sure that "on average" more bandwidth is available for the sessions changing their state from non-blocked to blocked. The parameter γ essentially controls the trade-off between the new and the ongoing session drop probabilities.

The authors in [25] developed an analytical model for the NR AP operation that supports bandwidth reservation. This approach is based upon an assumption that the session resource requirements do not vary much in the LoS non-blocked state as compared with the difference between the LoS blocked and the LoS non-blocked states. This allowed the authors to describe the evolution of the service process at the NR AP by a two-dimensional irreducible aperiodic Markov process, $\{n_0(t), n_1(t), t>0\}$, over the state-space

$$X^* = \{n_0 > 0, n_1 > 0 : n_0 d_0 + n_1 d_1 \le B\}, \tag{4.8}$$

where n_i ($i = 0, 1$) is the number of UEs in the LoS non-blocked and blocked states, correspondingly. The state spaces where drops of new (N) and ongoing (O) sessions may occur are specified as

$$\Pi_{N,0} = \{(n_0, n_1) : n_0 d_0 + n_1 d_1 + d_0 > \gamma B\},$$
$$\Pi_{N,1} = \{(n_0, n_1) : n_0 d_0 + n_1 d_1 + d_1 > \gamma B\}, \tag{4.9}$$
$$\Pi_O = \{(n_0, n_1) : n_0 d_0 + n_1 d_1 + d_0 - d_1 > B\},$$

where B is the total available bandwidth at the target NR AP.

The authors also studied the performance of the bandwidth reservation mechanism by relying on both user- and system-centric metrics. Particularly, as illustrated in Figure 4.11 where the average session rate is set to 50 Mbps, the use of bandwidth reservation does display the desired trade-off between the new and the ongoing session drop probabilities p_I and p_O, respectively. It is also demonstrated that this trade-off maintains across a wide range of session arrival intensities and depends on the average session rate. Importantly, reserving even a small fraction of the total bandwidth for the sessions already accepted by the system leads to a significant decrease in the ongoing session drop probability.

4.5.3 Summary

One may conclude that the use of bandwidth reservation at the mmWave APs allows for significant improvements in the ongoing session loss probability at the expense of some new session loss probability degradation. However, a particular balance between the new and the ongoing session drop probabilities as well as the associated value of the resource reservation coefficient depends heavily on the environmental characteristics, such as the density of users. At the initial 5G NR technology penetration phase, where the mmWave AP deployments are expected

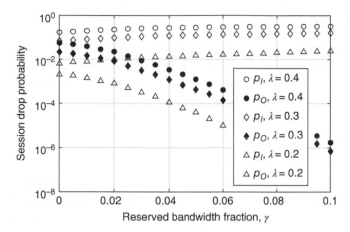

Figure 4.11 New and ongoing session drop probabilities as functions of reserved bandwidth.

to remain sparse, the proposed bandwidth reservation mechanism can be used as a standalone solution to improve session continuity of the rate-demanding IoT applications. At the later stages, it can be employed together with other techniques, such as MC capabilities [20], to equip the network operators with a powerful yet simple tool to manage the mmWave system performance.

4.6   Proactive Handover Mechanisms

4.6.1   Dynamic Blockage Avoidance

All of the techniques addressed so far to improve session continuity in 5G NR systems are inherently reactive, in the sense that they take measures when link blockage has already occurred. The common limitation of reactive MC and vertical handover solutions is that they introduce signaling overheads that cause delays in the service process, which may not be tolerated by the mission-critical IoT applications. While bandwidth reservation efficiently avoids these delays, its application is limited to those areas where the LoS blockage does not lead to outage. Hence, in many specific cases of interest proactive solutions that avoid blockage situations are preferred.

In completely random UE deployments, the application of proactive blockage avoidance techniques is extremely challenging [26]. One of the feasible methods to enable proactive switching in "purely" stochastic deployments is the use of machine learning techniques. The general approach here is to collect the LoS blockage and inter-blockage time statistics in real-time, and then utilize an appropriate machine learning method to decide upon the switching time instants. The relatively simple reinforcement learning and Markov decision process-based techniques can be employed to develop practical algorithms [27]. However, no solutions of this kind have been reported so far. Therefore, the performance of such proactive blockage avoidance techniques in stochastic environments has not yet been explored.

4.6.2 Deterministic AP Locations

A number of important 5G NR use cases are characterized by a certain regularity in the system and/or UE deployment. As an example, one may consider mmWave-based street deployments analyzed in detail in [14, 28], where the APs are located along the street on the building walls or lampposts with a fixed separation distance of d meters. The radius R_B denotes the distance where blockage leads to outage, while R_O is the distance corresponding to outage in non-blocked conditions. As one may observe, a deterministic AP deployment with the separation distance of d meters allows prediction of the points where AP switching needs to be performed. Although additional AP switches might be required at larger values of d as moving cars may lead to dynamic blockage, the knowledge of d is sufficient to predict the points where APs must be changed.

Based on the aforementioned principles, the authors in [28] developed a model for assessing the blockage-related metrics in 5G NR street deployments. Particularly, they considered a four-lane street where the tagged UE is associated with a vehicle moving in the inner lane and having a constant speed. The NR APs are assumed to be deployed at a certain height along the street with the constant inter-site distance (ISD) of d. Other vehicles are modeled as vertical screens with random lengths and heights, deployed according to a Poisson distribution with a certain intensity and moving at a constant speed. By varying the system parameters, the authors defined four distinct scenarios (highway, traffic jam, mission-critical road formation, and normal traffic conditions).

Among other metrics, the authors in [28] considered the intensity of session interruptions for the applications characterized by different outage tolerance times, see Figure 4.12 for $M = 3$. The delay-sensitive applications with $D_T = 0.1\,\text{s}$ experience the worst performance, while the minimal session interruption intensity is observed in the traffic jam scenario. The reason is that in the latter case the outage events are more seldom as compared with e.g. the highway scenario, while their durations are relatively long. Regular traffic conditions and mission-critical emergency scenarios are characterized by similar performance. Applications that may tolerate $D_T = 0.5\,\text{s}$ operate significantly better in highway environments as compared with other scenarios. The worst conditions for $D_T = 0.5\,\text{s}$ are observed for the traffic jam setup, while regular traffic conditions and mission-critical use cases remain in between these extremes.

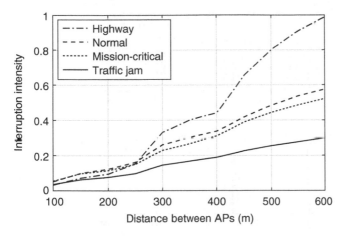

Figure 4.12 Intensity of session interruptions for considered scenarios.

4.6.3 Deterministic UE Locations/Trajectories

Similar to deterministic deployment of NR APs, predictability of the UE positions and/or movement also allows to develop efficient mechanisms for proactive blockage mitigation. As an example, one may consider NR-based edge-caching capable factory automation scenario for industrial IoT that was addressed in [29]. The authors considered a number of high-end IoT devices moving along well-defined trajectories at a constant speed, which are collaboratively involved into surveying and inspection operations. They analyzed three candidate data dissemination strategies: direct push, store-and-push, and forward-and-push. The first strategy relies on direct UE-AP communication, while the latter two employ UE caching capabilities.

Since the considered scenario is deterministic with respect to the UE positions, efficient algorithms can be developed to identify the time instants where blockage between the UE and the AP or between two UEs in question may occur. In particular, the authors in [29] proposed the use of two techniques: offline estimation and online statistical learning. The former approach is based on a combination of 3D modeling tools, geometric space transforms, and LoS mapping. As one may observe in Figure 4.13, the path diversity immediately available at each device allows to construct a reliable link to the AP at all times. As demonstrated by the authors, this diversity can be further exploited by a path selection module to drastically improve the performance of the system.

In the considered deployment, one may also utilize statistical learning approaches, where the IoT devices first spend some time to learn the blockage conditions and then push the learned data to the module responsible for the data dissemination strategy selection. Particularly, during the learning period, the IoT devices may collect information about their connection states with the AP and the neighboring

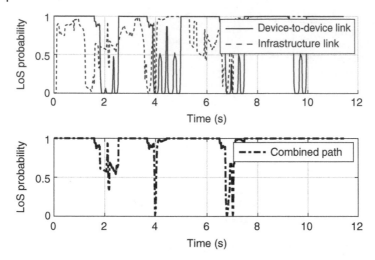

Figure 4.13 Example of LoS dependence on time.

UEs. Once such state information is obtained, it is propagated to the proximate IoT devices and the entire system transitions to its operating state. Once a change in the trajectory is enforced, the learning period starts all over again.

4.6.4 Summary

As compared with MC and bandwidth reservation techniques, proactive link selection currently remains at the relatively early stages of its development. However, these preliminary investigations promise impressive results in important realistic scenarios, thus calling for their in-depth study.

4.7 Conclusions

This chapter characterizes the effects of dynamic LoS blockage in mmWave 5G NR networks. The first part of this chapter offers a brief tutorial on modeling dynamic blockage processes in several representative 5G-grade consumer IoT scenarios that exploit NR connectivity. Particularly, it demonstrates that dynamic LoS path occlusions by moving objects can be modeled as a renewal process with the exponentially distributed duration of the non-blocked state and the generally distributed duration of the blocked state.

The chapter also highlights that in typical urban deployments the effects of dynamic blockage are essential in system modeling, since uninterrupted duration of the UE residing in the blocked state may reach several hundreds of milliseconds, which corresponds to tens of mmWave cellular superframes at

28 GHz. Further, it argues that a transition between the blocked and the non-blocked states occurs relatively frequently – on the order of seconds. Hence, dynamic blockage introduces severe fluctuations in the amount of occupied radio resources, which deserves focused attention in 5G network design, modeling, and deployment.

In the second part of this chapter, the state-of-the-art solutions to improve session continuity in the presence of dynamic blockage are introduced and discussed in the order of maturity of their potential implementation: from current 3GPP standardization efforts to conceptual ideas. It is particularly highlighted that 3GPP MC is an efficient technique to combat blockage when handling a large number of high-rate IoT sessions. For a typical urban deployment, it is shown that maintaining four simultaneous links is sufficient to reduce the UE outage probability by a half, while higher degrees of MC do not result in notable performance gains.

Out of many MC policies – algorithms to reroute data transmissions in case of a sudden link blockage – those jointly considering the link blockage probability and the achievable link throughput are confirmed to be the most efficient and robust. Later, it is also demonstrated that even when an IoT device does not support MC, the network can still address the dynamic blockage problem partially by exploiting the concept of bandwidth reservation. It is made clear that by reserving only 5% of its radio resources, the system can decrease the session drop probability by up to several orders of magnitude.

Finally, the concept of proactive link rerouting to lower the chances of dynamic blockage is outlined. It is suggested that in, e.g. factory automation scenarios a combined use of infrastructure and device-to-device connections allows for achieving higher reliability. This concept is especially beneficial in (semi-)deterministic deployments, where node mobility is more predictable. In summary, the present chapter is dedicated to dynamic blockage effects and their impact on the prospective mmWave systems, which deserve careful consideration across a wide range of 5G-grade IoT scenarios. Summarized mathematical tools and numerical findings can contribute to further development of reliable mmWave connectivity for multiple 5G verticals: from wearable AR/VR systems to factory automation and connected (semi-)autonomous vehicles.

References

1 3GPP (2018). NR; NR and NG-RAN overall description (Release 15). 3GPP TS 38.300 V15.2.0, June 2018.

2 MacCartney, G.R., Rappaport, T.S., and Rangan, S. (2017). Rapid fading due to human blockage in pedestrian crowds at 5G millimeter-wave frequencies. *Proceedings of IEEE Global Communications Conference (GLOBECOM)* (December 2017), pp. 1–7.

3 Rappaport, T.S., MacCartney, G.R., Samimi, M.K., and Sun, S. (2015). Wideband millimeter-wave propagation measurements and channel models for future wireless communication system design. *IEEE Transactions on Communications* 63 (9): 3029–3056.

4 Gapeyenko, M., Petrov, V., Moltchanov, D. et al. (2018). Flexible and reliable UAV-assisted backhaul operation in 5G mmWave cellular networks. *IEEE Journal on Selected Areas in Communications* 36 (11): 2486–2496.

5 3GPP (2018). Study on channel model for frequencies from 0.5 to 100 GHz (Release 15). 3GPP TR 38.901 V15.0.0, June 2018.

6 Haneda, K., Heino, M., and Jarvelainen, J. (2018). Total array gains of millimeter-wave mobile phone antennas under practical conditions. *Proceedings of IEEE 87th Vehicular Technology Conference (VTC Spring)* (May 2018), pp. 1–6.

7 Haneda, K., Zhang, J., Tan, L. et al. (2016). 5G 3GPP-like channel models for outdoor urban microcellular and macrocellular environments. *Proceedings of IEEE 83rd Vehicular Technology Conference (VTC Spring)* (May 2016), pp. 1–7.

8 Jacob, M., Priebe, S., Kürner, T. et al. (2013). Fundamental analyses of 60 GHz human blockage. *Proceedings of European Conference on Antennas and Propagation (EuCAP)* (April 2013), pp. 117–121.

9 Wang, Y., Venugopal, K., Molisch, A.F., and Heath, R.W. (2017). Blockage and coverage analysis with mmWave cross street BSs near urban intersections. *Proceedings of IEEE International Conference on Communications (ICC)* (May 2017), pp. 1–6.

10 Gapeyenko, M., Samuylov, A., Gerasimenko, M. et al. (2017). On the temporal effects of mobile blockers in urban millimeter-wave cellular scenarios. *IEEE Transactions on Vehicular Technology* 66 (11): 10124–10138.

11 V. Petrov, K. Mikhaylov, D. Moltchanov et al. When IoT keeps people in the loop: a path towards a new global utility. *IEEE Communications Magazine*, 57(1):114–121, 2019.

12 Gapeyenko, M., Samuylov, A., Gerasimenko, M. et al. (2016). Analysis of human-body blockage in urban millimeter-wave cellular communications. *Proceedings of IEEE International Conference on Communications (ICC)* (May 2016), pp. 1–7.

13 Samuylov, A., Gapeyenko, M., Moltchanov, D. et al. (2016). Characterizing spatial correlation of blockage statistics in urban mmWave systems. *Proceedings. of IEEE Global Communications Workshops (GLOBECOM)* (December 2016), pp. 1–7.

14 Petrov, V., Lema, M.A., Gapeyenko, M. et al. (2018). Achieving end-to-end reliability of mission-critical traffic in softwarized 5G networks. *IEEE Journal on Selected Areas in Communications* 36 (3): 485–501.

15 Andreev, S., Petrov, V., Dohler, M., and Yanikomeroglu, H. (2019). Future of ultra-dense networks beyond 5G: Harnessing heterogeneous moving cells. *IEEE Communications Magazine* 57 (6): 86–92.

16 Ross, S.M. and Seshadri, S. (1999). Hitting time in an M/G/1 queue. *Journal of Applied Probability* 36 (3): 934–940.

17 3GPP (2018). NR; Multi-connectivity; Overall description; Stage-2. 3GPP TS 37.340 V15.2.0, June 2018.

18 Giordani, M., Mezzavilla, M., Rangan, S., and Zorzi, M. (2016). Multi-connectivity in 5G mmWave cellular networks. *Proceedings of Mediterranean Ad Hoc Networking Workshop* (June 2016), pp. 1–7.

19 Drago, M., Azzino, T., Polese, M. et al. (2018). Reliable video streaming over mmwave with multi connectivity and network coding. *Proceedings of International Conference on Computing, Networking and Communications (ICNC)* (March 2018), pp. 508–512.

20 Petrov, V., Solomitckii, D., Samuylov, A. et al. (2017). Dynamic multi-connectivity performance in ultra-dense urban mmWave deployments. *IEEE Journal on Selected Areas in Communications* 35 (9): 2038–2055.

21 Tesema, F.B., Awada, A., Viering, I. et al. (2015). Mobility modeling and performance evaluation of multi-connectivity in 5G intra-frequency networks. *Proceedings of IEEE Globecom Workshops (GLOBECOM)* (December 2015), pp. 1–6.

22 Gapeyenko, M., Petrov, V., Moltchanov, D. et al. (2019). On the degree of multi-connectivity in 5G millimeter-wave cellular urban deployments. *IEEE Transactions on Vehicular Technology* 68 (2): 1973–1978.

23 Oehmann, D., Awada, A., Viering, I. et al. (2018). Modeling and analysis of intra-frequency multi-connectivity for high availability in 5G. *Proceedings of IEEE 87th Vehicular Technology Conference (VTC Spring)* (June 2018), pp. 1–7.

24 Seitz, N. (2003). ITU-T QoS standards for IP-based networks. *IEEE Communications Magazine* 41 (6): 82–89.

25 Moltchanov, D., Samuylov, A., Petrov, V. et al. (2019). Improving session continuity with bandwidth reservation in mmWave communications. *IEEE Wireless Communications Letters* 8 (1): 105–108.

26 Petrov, V., Moltchanov, D., and Koucheryavy, Y. (2016). Applicability assessment of terahertz information showers for next-generation wireless networks. *Proceedings of IEEE International Conference on Communications (ICC)* (May 2016), pp. 1–7.

27 Sutton, R.S. and Barto, A.G. (1998). *Introduction to Reinforcement Learning*, vol. 135. Cambridge, MA: MIT Press.

28 Begishev, V., Samuylov, A., Moltchanov, D. et al. (2018). Connectivity properties of vehicles in street deployment of 3GPP NR systems. *Proceedings of IEEE Global Communications Conference (GLOBECOM)* (December 2018), pp. 1–7.

29 Orsino, A., Kovalchukov, R., Samuylov, A. et al. (2018). Caching-aided collaborative D2D operation for predictive data dissemination in industrial IoT. *IEEE Wireless Communications* 25 (3): 50–57.

5

Radio Resource Management Techniques for 5G Verticals

S.M. Ahsan Kazmi[1], Tri Nguyen Dang[2], Nguyen H. Tran[3], Mehdi Bennis[4], and Choong Seon Hong[2]

[1] *Institute of Information Security and Cyber Physical Systems, Innopolis University, Innopolis, Russia*
[2] *Department of Computer Science and Engineering, Kyung Hee University, Seoul, Korea*
[3] *School of Computer Science, The University of Sydney, Sydney, NSW, Australia*
[4] *Department of Communications, University of Oulu, Oulu, Finland*

Abstract

A cellular communication network can broadly be categorized into two parts, the core network and the radio access network (RAN). This chapter focuses on the latter part of cellular communication network, i.e. RAN. It briefly presents 5G specifications along with the enabling technologies that will support bringing the 5G networks to fruition. Then the chapter discusses different types of resources available for the RAN of future cellular systems that need to be managed for effective operations. It also discusses network slicing which can be considered as a promising scheme to meet the heterogeneous requirements produced by various 5G verticals. The chapter further provides a use-case that focuses on one of the most important requirements of virtual reality (VR), i.e. offloading computation task. Offloading intensive tasks to more resourceful devices such as clouds or mobile edge computing servers increases the computational capacity of VR devices.

Keywords *5G verticals; cellular systems; network slicing; offloading computation task; radio access network management; radio resource management techniques; virtual reality*

5G Verticals: Customizing Applications, Technologies and Deployment Techniques, First Edition.
Edited by Rath Vannithamby and Anthony C.K. Soong.
© 2020 John Wiley & Sons Ltd. Published 2020 by John Wiley & Sons Ltd.

5.1 Introduction

Massive advancements in wireless networks have helped us in optimizing our operations, economics, health, and all modern industries. Among wireless networks, the cellular networks are the most popular [1]. A cellular communication network can broadly be categorized into two parts, the core network (CN) and the radio access network (RAN). This chapter focuses on the latter part of cellular communication network, i.e. RAN. A *RAN* is an integral part of the cellular communication system that implements the *radio access technology* (RAT) to enable the subscribers to access the CN. A typical RAN consists of both wireless and wired links along with control or switching sites [2]. In a cellular system, the limited resources must be shared effectively among the its subscriber to enhance the overall quality of service (QoS) of the network. In the following section, we briefly present 5G specifications along with the enabling technologies that will support bringing the 5G networks to fruition. Then we discuss different types of resources available for the RAN of future cellular systems that need to be managed for effective operations.

5.2 5G Goals

According to [3], there will be a 33-fold increase in mobile traffic worldwide in 2020 compared with 2010. To cope with the gigantic increase in mobile data traffic indicated by the statistics, effective planning of the network is imperative. The 5G standard in line with the trend of growth in traffic is providing guidelines for future cellular networks [4]. The 5G cellular networks are intended to provide peak data rates of up to 10 Gbps, latency up to 1 ms, 1000 times more devices, 10-fold energy efficiency, and high reliability according to METIS.[1] As a result of massive technological revolutions, the demands posed by end users have increased drastically. To meet these demands, ITU has classified the future 5G services into three main categories: ultra-reliable and low-latency communications (URLLC), enhanced mobile broadband (eMBB), and massive machine type communication (mMTC) services. The existing mobile network architecture was designed to meet requirements for voice and conventional mobile broadband (MBB) services. Furthermore, the previous cellular generations were primarily designed to only fulfill the human communication requirements such as voice and data. However, 5G networks are expected to facilitate industrial communication as well in order to grow industry digitalization. This will enable innovative services and networking capabilities for new industry stakeholders. The 5G technology is expected to provide connectivity and communication needs with specific solutions to vertical sectors such as automotive, health care, manufacturing, entertainment

---

1 https://bwn.ece.gatech.edu/5G_systems.

and others in a cost-effective manner. Note that these novel services have very diverse requirements, thus, having traditional RAN and core management solutions for every service cannot guarantee the end user QoS. Moreover, some incremental improvement has been observed due to installation of small cells under macro cell coverage (i.e. heterogeneous networks [HetNets]) especially in congested locations. The concept of HetNets has already been implemented in current networks.[2] Other promising approaches for enabling the future RANs include enabling device-to-device (D2D) communication to reduce network traffic [5, 6], installing cache storage at the access networks to reduce delays [7], performing computation at the base stations (BSs) for real time analytics, and allowing the use of unlicensed spectrum such as LTE-Unlicensed (LTE-U) to further enhance the network capacity.

5.3 Radio Access Network Management

Management of resources in future RANs will become one of the biggest and crucial challenges for efficient performance of the complete network. Furthermore, the future RAN will be a very complex mixture of heterogeneous coexisting technologies that would require tight coordination for utilizing the shared and limited resources [8]. The resources in a 5G RAN can be broadly categorized into three categories: (i) spectrum and power resources available in the radio access terminal for carrying out communication, (ii) cache resources available for storage in the BS to reduce access delays, and (iii) edge computing resources available at the access networks for performing real time computation.

The biggest challenge pertains to the limited available wireless spectrum which falls in the spectrum resource management of RANs. The spectrum suffers from congestion due to the exponential increase in the number of wireless devices and bandwidth hungry applications [9]. To tackle the issue of spectrum scarcity, researchers are trying to explore new high radio frequency bands. In 5G, cellular network technology is intending to use the millimeter wave (mmWave) frequencies band to enable higher data rates [10]. New spectrum above 6 GHz is considered by 5G New Radio (NR) for addressing specific use cases requiring extremely high data rates. Note that bands below 6 GHz are crucial to support wide-area coverage and 5G scenarios of mMTC. The 5G NR would be capable of operating on multiple bands to address diversified requirements from the envisioned 5G usage scenarios. Moreover, power management on these spectrums will also play a very crucial role for enhancing the performance. Therefore, the main goal is to develop efficient RAN schemes for spectrum and power allocation that can efficiently perform resource management [11]. Next, we provide an overview of some recent advances in radio resource management approaches for the RAN.

---

2 https://www.3gpp.org/technologies/keywords-acronyms/1576-hetnet.

The problem of spectrum resource allocation in two-tier HetNets is considered in [12]. In resource allocation, macro-cell protection is achieved through use of constraints on cross-tier interference. The resource allocation problem is formulated as a *mixed-integer resource allocation problem* which is solved using two different algorithms. One algorithm is based on a duality-based optimization approach and the other is based on matching game theory. However, this work only considered subchannel allocation and assumed uniform power allocation on each subchannel. In [13], the authors considered a two-tier network consisting of macro cells and femto cells with co-channel assignments. They formulated a joint resources allocation and power control optimization problem which is of mixed integer non-linear programming (MINLP) type with the objective of minimizing the total transmit power. The proposed MINLP problem is transformed using a reformation-linearization technique and then a QoS-based radio resource management algorithm is proposed to minimize the interference and improve QoS.

Caching is used to store the popular contents temporarily in a cache to enable lower latency and reduced back haul usage. According to statistics, video streaming traffic aids a significant portion (54%) of the total Internet traffic which is expected to grow to 71% in 2019 [14]. Most often the users request similar video contents which induces overhead on the back-haul networks. Caching the content at the BS will reduce the congestion of the back-haul links and latency experienced by the user. Typically, the architecture of cache enabled cellular networks can be divided into four categories, i.e. *cache enabled C-RAN*, *cache enabled macro-cellular networks*, *cache enabled D2D networks*, and *cache enabled HetNets* [15]. Note that the cache space is very limited compared with the number of contents. Thus, efficient caching schemes for saving popular contents that can be reused later with minimum delay need to be deigned. Thus, efficient caching schemes are required for 5G RANs that will play a vital role in reducing the access delay and saving back-haul bandwidth. Next, we discuss some recent work on caching schemes.

Yang et al. [16] considered a three-tier network consisting of D2D pairs, relays, and BSs. The node (uses, relays, and BSs) locations are first modeled as mutually independent Poisson point processes (PPPs) and then protocol for content access is proposed. Along with this, the authors derived analytical expressions for outage probability and average ergodic rate for subscribers considering different cases. Further, derivation of delay and throughput using continuous-time Markov process and multiclass processor-sharing model are also performed. The authors considered unicast point-to-point transmission, which minimize the backhaul traffic but fails to reduce "on air" congestion effectively. Along with this, the wireless medium broadcast nature in contrast to wired medium is not effectively exploited. In [17] energy efficient caching is proposed to jointly optimize the cache placement and cache hit ratio. A technique using short noise model (SNM) is proposed to estimate the content popularity. The cache hit rate is then improved by proposing distributed caching policy. Along with this, the authors have formulated the caching problem as an optimization problem and presented a closed-form solution that maximizes the energy efficiency for the optimal cache capacity.

The 5G network will be required to perform a number of computation tasks especially after the advent of the Internet of things (IoT). Efficient computation will play a vital role in realizing a number of real time services. One option is to adopt cloud computing for task offloading. Cloud computing enhances the user's overall quality of experience by providing the shared computing and storage resources online as a service in an elastic, sustainable, and reliable manner. Although cloud computing offers significant advantages, latency sensitive applications such as self-driving cars, mission critical applications, industrial automation, and augmented reality, etc. incur performance degradation because of the distant locations of the cloud. Edge computing is a solution to delay sensitive applications that push the computing resources to the edge of the network. In [18], Satyanarayanan et al. presented the concept of cloud-let, a small-scale data center positioned at the network edge to enable the execution of resource-intensive applications with low latency. In computing [19], Cisco coined the term Fog computing, an architecture based on usage of edge devices to enable local computing resources. Mobile edge computing (MEC) was the term coined by the European Telecommunications Standards Institute (ETSI) [20] to allow the placement of storage and computing resources at the BS. All these aforementioned technologies aim to reduce the computational time delay and achieve the required QoS for the specific 5G use case.

In [21] and [22], the authors have used the binary task offloading approach. The authors in [21] considered a single user MEC system with energy harvesting. The authors defined an execution cost function which is the weighted sum of the task dropping cost and the execution delay. The dropping cost reflects the tasks that are dropped either due to deep fading or due to lack of energy at the mobile device, which is only powered by the harvested energy in the system model. They formulated an optimization problem to minimize the time average of the execution cost. Further, they have proposed an online Lyapunov optimization-based approach to solve the optimization problem. The main disadvantage of the proposed system is the use of only a harvesting source for powering the mobile device. The work in [22] considers a fleet of drones (user's nodes) running resource-intensive tasks such as identification and classification of objects. A cost function to jointly minimize the delay and energy consumption is defined for which a game theory based distributed offloading scheme is proposed. The theoretical game with three different strategies (such as local task computation, task offloading to a BS through a local wireless network, and offloading of task to a sever through a cellular network) and drones as players is considered to perform task offloading. The proposed approach has the advantage of its distributed nature; however, its associated complexity is not discussed.

To fulfill these diverse heterogeneous requirements posed by vertical industries, a novel and promising concept of network slicing can be adopted.

5.4 Network Slicing

Network slicing has recently attracted significant attention due to its wide applicability for 5G networks [23]. Through network slicing, the physical infrastructure of a cellular operator can be logically divided into slices of heterogeneous capabilities that can support various services with different requirements [24,25]. For example, an operator can provide a dedicated slice of a specific network's capabilities for handling augmented reality applications with URLLC and another dedicated slice with different network capabilities for video-on-demand services that require high throughput. This enables network operators to provide network as a service and enhances the network efficiency. One important aspect in enabling network slicing is to enforce strong isolation between different slices such that any actions in one slice do not affect other operating slices.

Network slicing is also being supported by 3GPP by defining a novel network architecture that supports slicing. In particular, the 3GPP working group SA2 has already defined the basis for building an evolved CN infrastructure managing multiple slices on the same network infrastructure [26]. Network slicing can be easily realized due to the revolutionary technologies of software defined networking (SDN) and network function virtualization (NFV). The work in [27] presents a detailed survey on wireless resource slicing and the challenges associated with isolation of wireless slices. Other notable work relating to novel network slicing architectures can also be found [28–30]. All these aforementioned works basically discuss how to enable network slicing and its benefits in 5G networks. Specifically, they discuss the roles and interactions of network slicing enablers such as SDN and NFV in the 5G network architecture. They do not consider specific algorithmic approaches for slice creation, slice allocation, slice interactions, admission control, etc. which are equally important to bring network slicing to fruition. Moreover, considering the potential benefits of network slicing, a number of mobile operators have shown keen interest and efforts in adopting network slicing in their network. SK Telecom and Ericsson successfully demonstrated network slice creation and operation for augmented reality solutions in 2015. Similarly, dynamic network slicing proof of concept for 5G CNs was also provided in 2016 by Ericsson and NTT DoCoMo. Other notable efforts pertaining to network slicing have been demonstrated by operators such as Huawei, Deutsche Telekom, ZTE, and China Mobile.

Even though there is technology development on network slicing, the majority of it is focused on slicing at the CN due to massive advancements in SDN, virtualization, and NVF technologies. On the other hand, there are limited researches on RAN slicing, with most of them focused on slicing a single RAN resource, e.g. either spectrum or BS [28–31]. Note that network capacity is inarguably a critical resource of RAN, however, other resources of RAN such as cache space, backhaul capacity and computing at the RAN also need to considered for RAN network slicing. The work in [32] considers caching and back haul limitation of 5G RANs.

5.5 Use Case: Virtual Reality

Virtual reality (VR) is an interactive digital real time experience taking place with a simulated artificial environment. It mainly includes audio and visual interactions. Its goal is to generate a real time virtual environment that mimics the human perceptions. To support this real time virtual environment, the wireless system needs to cope with a massive amount of bandwidth and latency requirements which was not possible with the previous cellular generations. Indeed, some VR technologies such as VR goggles are already emerging, however, the full potential of VR systems to achieve a fully immersive experience is yet to be explored. A detailed work on VR requirements, its enablers and its challenges are discussed in [33]. In this use case, we focus on one of the most important requirements of VR, i.e. computation offloading. Offloading intensive tasks to more resourceful devices such as clouds or MEC servers is an effective option to increase the computational capacity of VR devices.

5.5.1 System Model

Let $\mathcal{N} = \{1,...,N\}$ be the set of N mobile users with VR capability as shown in Figure 5.1, and $\mathcal{M} = \{1, ..., M\}$ be the set of M MEC servers. Each mobile user $i \in \mathcal{N}$ has a computational task to offload that is denoted by $\tau_i = \{S_i, T_i, C_i\}$, where T is the worst case execution time of the task, C is the budget of user i for offloading the task, and Si is size of the task. Moreover, we assume the task to be a video task. Typically, in a video frame, there are four components: horizontal pixels, vertical pixels, frame rate, and color depth, therefore, we consider

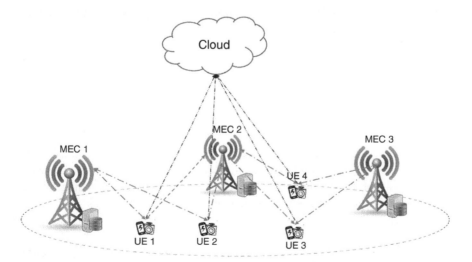

Figure 5.1 System model.

$S_i = \{H_i, V_i, F_i, L_i\}$, where H is horizontal pixels, V is vertical pixels, F is number of frames per second, and L is the length of video. Thus, the size of video is calculated using the following equation:

$$S_i = \frac{H \times V \times F \times L \times 3}{1024 \times 1024} \tag{5.1}$$

In this work, we assume that the bandwidth allocated for user i from different MEC j is different, $d_{i,j} \neq d_{i,k}, \forall i \in \mathcal{N}, \forall j,k \in \mathcal{M}, j \neq k$. This difference depends upon factors such as location, distance and power level of devices. In our model, we consider two main factors pertaining to a user that affects the offloading decision: processing cost of the task and the transmission time of the task. Thus, we have two cases for any user $i \in \mathcal{N}$. The first case occurs when user i directly uploads to the cloud server. In this case, let c_{i0} be the cost for data storage where c_{i0} is calculated using a linear function with coefficient δ. The goal of assuming a linear cost function is to enable the pay-as-you-go policy. This means that the more data uploaded, the greater the cost. The second case occurs when user i uploads the task to an MEC server. Note that each MEC server has a limit on its processing capacity which is denoted by Γ_j. Now, we define a bandwidth allocation matrix for the uplink transmission:

$$\mathbf{U} = \begin{bmatrix} u_{11} & \cdots & u_{1M} \\ \vdots & \ddots & \vdots \\ u_{N1} & \cdots & u_{NM} \end{bmatrix} \tag{5.2}$$

where u_{ij} is the uplink bandwidth allocated for user i when offloading to MEC j. Given the bandwidth and size of the video frames x_{ij}; the uplink transmission time t_{ij}^u can be calculated as follows:

$$t_{ij}^u = \frac{x_{ij}}{u_{ij}}. \tag{5.3}$$

And the processing cost is:

$$c_{ij} = \delta_j \times x_{ij}. \forall j \in \mathcal{M} \tag{5.4}$$

Similar to the cloud case, here also we use the linear cost function for calculating the processing cost in order to follow the pay-as-you-go policy and δ represents the coefficients of the cost function. Then, the bandwidth allocation matrix for the downlink transmission can be represented by:

$$\mathbf{D} = \begin{bmatrix} d_{11} & \cdots & d_{1M} \\ \vdots & \ddots & \vdots \\ d_{N1} & \cdots & d_{NM} \end{bmatrix} \tag{5.5}$$

where u_{ij} is the downlink bandwidth allocated for user i when associated with MEC j. Given the size of output S'_i from MEC j, the downlink transmission time can be defined as follows:

$$t^d_{ji} = \frac{\eta x_{ij}}{d_{ij}},$$

(5.6)

where η is the coefficient representing the relation between input and output data. For example: $\eta = 0{:}5$ means that the result is equal to 50% of input data. Then, the total transmission cost is defined as follows:

$$t_{ij} = \left(t^u_{ij} + t^d_{ji}\right)$$

(5.7)

Note that the transmitted and received time for a user i from the central cloud server is denoted by t^u_{i0} and t^d_{0i}, respectively. Moreover, we assume that the demand of users is to be served at the MEC for low processing time given it has enough capability. Therefore, to minimize the total cost of the end user, we normalize the total transmission and processing cost functions by dividing by the original cost, i.e. cost required for a user i when uploading its task to the cloud server.

$$t'_i = \left(\frac{\sum\limits_{j \in M} t^u_{ij}}{t^u_{i0}} + \frac{\sum\limits_{j \in M} t^d_{ji}}{t^d_{i0}}\right), \forall i \in N$$

(5.8)

and the processing cost

$$c'_i = \frac{\sum\limits_{j \in M} c_{ij}}{c_{i0}}. \forall i \in N$$

(5.9)

5.5.2 Problem Formulation

In this subsection, we formulate our problem of task offloading for the VR enabled users. We aim to minimize the total transmission and processing cost of the network. Then, our problem can be stated as follows:

$$\underset{x}{minimize} : \sum\limits_{i \in N} \alpha \left(\frac{\sum\limits_{j \in M} t^u_{ij}}{t^u_{i0}} + \frac{\sum\limits_{j \in M} t^d_{ji}}{t^d_{i0}}\right) + (1 - \alpha) \frac{\sum\limits_{j \in M} c_{ij}}{c_{i0}}$$

(5.10)

$$subject\ to : \sum\limits_{j \in M} x_{ij} = \frac{H_i \times V_i \times F_i \times L_i \times 3}{1024 \times 1024}, \forall i \in N,$$

(5.11)

$$\sum_{i \in N} x_{ij} \leq \Gamma_j, \forall j \in M, \tag{5.12}$$

$$\sum_{j \in M} t_i \leq T_i, \forall i \in N, \tag{5.13}$$

$$\sum_{j \in M} c_i \leq C_i, \forall i \in N, \tag{5.14}$$

$$t_i' < 1, \forall i \in N, \tag{5.15}$$

$$c_i' < 1, \forall i \in N, \tag{5.16}$$

$$x_{ij} \geq 0, \forall i \in N, \forall j \in M. \tag{5.17}$$

Our objective function Eq. (5.10) considers both the transmission and processing cost in which α represents the trade-off coefficient that ranges from [0, 1]. Constraint Eq. (5.11) guarantees that all demand is served by the network, whereas the constraint Eq. (5.12) represents that the MEC capacity is not violated. Furthermore, the constraint Eq. (5.13) states that the budget of any user does not exceed the processing cost. Lastly, constraint Eqs. (5.15) and (5.16) state that the cost of transmission, and processing at MEC are less compared with that of a central cloud, respectively. To solve this problem we use the alternating direction method of multipliers (ADMM) approach. ADMM has the capability of parallel solving the designed problem in a distributed fashion in which each user or MEC server will solve its individual variable(s). Next, we describe our designed approach.

5.5.3 ADMM-Based Solution

Here, we redefine the objective function Eq. (5.10) as follows:

$$(5.10) = \alpha \left(\sum_{j \in M} \left(\frac{x_{ij}}{u_{ij} t_{i0}^u} + \frac{\eta x_{ij}}{d_{ij} t_{ji}^d} \right) \right) + (1 - \alpha) \left(\frac{\sum_{j \in M} \delta_j x_{ij}}{c_{i0}} \right) \tag{5.18}$$

$$= \alpha \left(\sum_{j \in M} \left(\frac{1}{u_{ij} t_{i0}^u} + \frac{\eta}{d_{ij} t_{ji}^d} \right) \right) x_{ij} + (1 - \alpha) \left(\frac{\sum_{j \in M} \delta_j}{c_{i0}} \right) x_{ij} \tag{5.19}$$

$$= \left(\alpha \sum_{j \in M} \left(\left(\frac{1}{u_{ij} t_{i0}^u} + \frac{\eta}{d_{ij} t_{ji}^d} \right) + (1-\alpha) \left(\frac{\sum_{j \in M} \delta_j}{c_{i0}} \right) \right) \right) x_{ij} \tag{5.20}$$

$$= f_i(\mathbf{x}_i) \tag{5.21}$$

where $\mathbf{x}_i \triangleq \{x_{ij}, j \in M\}$ then the optimization problem can be rewritten as follows:

$$\underset{x}{\text{minimize}} : \sum_{i \in N} f_i(\mathbf{x}_i) \tag{5.22}$$

$$\text{subject to} : \mathbf{1}^T \mathbf{x}_i = S_i, \forall i \in N \tag{5.23}$$

$$\mathbf{1}^T \mathbf{x}_j \leq \Gamma_j, \forall j \in M \tag{5.24}$$

$$\sum_{j \in M} t_i \leq T_i, \forall i \in N \tag{5.25}$$

$$\sum_{j \in M} c_i \leq C_i, \forall i \in N \tag{5.26}$$

$$t_i' < 1, \forall i \in N \tag{5.27}$$

$$c_i' < 1, \forall i \in N \tag{5.28}$$

$$x_{ij} \geq 0, \forall i \in N, \forall j \in M \tag{5.29}$$

We then define the feasible set for the problem Eq. (5.22) as follows:

$$\mathcal{X} \triangleq \{ \mathbf{x}_i | (1.23), (1.24), (1.25), (1.26), (1.27), (1.28), (1.29) \} \tag{5.30}$$

Then, following the ADMM framework, we introduce a new variable z such that

$$\begin{aligned} \underset{x}{\text{minimize}} : \quad & \sum_{i \in N} f_i(\mathbf{x}_i) + h(z) \\ \text{subject to} : \quad & \mathbf{x}_i = z \\ & \mathbf{x}_i \in \mathcal{X}, \forall i \in N \end{aligned} \tag{5.31}$$

Where $h(z) = 0$ when $\mathbf{x}_i \in \mathcal{X}$.

$$h(z) = I_{\mathcal{X}}(z) = \begin{cases} 0, & \mathbf{x}_i \in \mathcal{X} \\ \infty, & \text{otherwise} \end{cases} \tag{5.32}$$

Then, the augmented Lagrangian function of Eq. (5.31) is as follows:

$$\mathcal{L}(\mathbf{x}, z, \lambda) = \sum_{i \in N} \left(f_i(\mathbf{x}_i) + \lambda_i^T (x_i - z) + \frac{\rho}{2} \| x_i - z \|_2^2 \right) \tag{5.33}$$

Based on the solution from [34], the resulting ADMM variable updates are the following:

$$x_i^{k+1} = \arg\min \left(f_i(x_i) + \lambda_i^{kT} (x_i - z^k) + \frac{\rho}{2} \| x_i - z^k \|_2^2 \right) \tag{5.34}$$

$$z^{k+1} = \arg\min \left(h(z) + \sum_{i=1}^{N} \left(-\lambda_i^{kT} z + \frac{\rho}{2} \| x_i^{k+1} - z \|_2^2 \right) \right) \tag{5.35}$$

$$\lambda_i^{k+1} = \lambda_i^k + \rho \left(x_i^{k+1} - z^{k+1} \right) \tag{5.36}$$

Algorithm 5.1 represents the pseudo code of our approach. Initially, we input the set of users, MEC servers, and its downlink and uplink allocation capabilities. Then, we initialize all the variables for the first iteration and fix the maximum number of iterations to 1000 runs. Moreover, the Lagrangian penalty term and trade-off coefficient is set to 0.5. For each iteration, each user i updates its offloading decision. Once all users have updated their respective offloading decision, the MEC updates the Lagrange multiplier variable. This is followed by updating the objective function 1.10 based on the current iteration. After limited iterations, the solution converges to an optimal value.

Algorithm 5.1   ADMM-Based Task Offloading

---

1) **Input**: Initialization for $\mathcal{N}, \mathcal{M}, \mathbf{D}, \mathbf{U}$
2) **Output**: Minimal offloading cost
3) Initialization
4) max_iteration $= 1000, \rho = 0.5, \alpha = 0.5 \, \mathbf{x}_i^0 \geq 0, \lambda_i^0 \geq 0, z \geq 0, t_{i0}^u, t_{i0}^d, c_{i0}, \forall i \in N$
5) **for** $k \in$ max_iteration **do**
6) Each user $i \in \mathcal{N}$ updates its offloading decision by Eqs. (5.34), (5.35), and (5.32), respectively, parallelly
7) After getting updated values from all users each MEC will update λ using Eq. (5.36), in parallell
8) After all variables updated, update the objective function Eq. (5.10)
9) **end for return** Optimal value of objective Eq. (5.10)

5.5.4 Performance Analysis

In our simulation, the MECs are assumed to be deployed at a fixed location, and N mobile users are deployed following a homogeneous PPP where each user has a single task to offload that can be separable. The input parameters of downlink and uplink bandwidth allocation follow a homogeneous uniform distribution ranging from 0 to 5. In our simulations, the η value is varied from 0.4 to 0.7. Note that all statistical results are averaged over a hundred simulation runs of random location of users and bandwidth allocation. Moreover, we compare the performance of our scheme (i.e. ADMM-based approach) with three other schemes. The first scheme uses a centralized method in which the optimal solution is calculated using the "convex.JL". We represent this scheme as the "Centralized" scheme. Then, we calculate the solution using a greedy approach in which the best first search approach is used to select the offloading decision of user to MEC. This scheme is represented as the "Greedy" scheme. Finally, the last scheme is the "Random" scheme in which we pick random MECs for offloading its task with uniform distribution.

Figure 5.2 represents the average utility obtained by all the schemes by increasing the number of MECs in the network. It can be seen that the average utility increases with number of MEC servers. Moreover, the performance of the proposed scheme is significantly higher than the greedy and random schemes. Furthermore, the performance of the proposed ADMM scheme is indifferentiable from the centralized scheme, i.e. optimal solution under all scenarios. Similarly, Figure 5.3 represents the average processing cost obtained by all the schemes by increasing the number of MECs in the network. It can be inferred that the processing cost decreases as the number of MEC servers increases in the network as portions of task are divided among multiple MEC servers. Similarly, the performance of the ADMM scheme is similar to the centralized scheme, thus, obtaining an optimal solution.

5.6 Summary

Radio resource management is categorized among the biggest challenge for the 5G networks due to the proliferation of heterogeneous devices. Moreover, the introduction of 5G verticals and the need to fulfill heterogeneous stringent requirements based on novel applications further complicates the radio resource management process. In this chapter, we presented an overview of radio resources available in RAN and some recent approaches to manage the network. Moreover, we also discussed network slicing which can be considered as a promising scheme to meet these heterogeneous requirements produced by various 5G verticals.

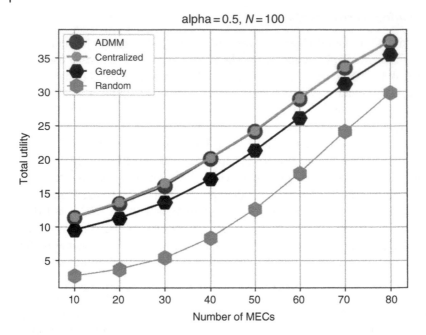

Figure 5.2 Average utility versus number of MEC servers.

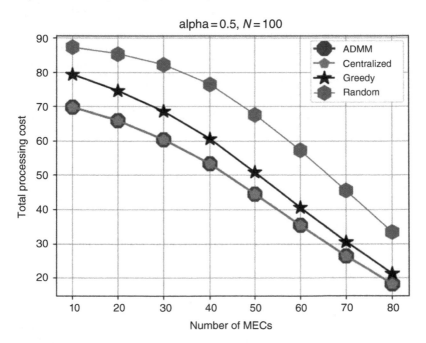

Figure 5.3 Average processing cost versus number of MEC servers.

Finally, we provided a use-case that focuses on one of the most important requirements of VR, i.e. offloading computation task. Offloading intensive tasks to more resourceful devices such as clouds or MEC servers increases the computational capacity of VR devices.

References

1 Hong, C.S., Ahsan Kazmi, S.M., Moon, S. et al. (2016). SDN based wireless heterogeneous network management. *AETA 2015: Recent Advances in Electrical Engineering and Related Sciences*, pp. 3–12. Springer.

2 Raza, H. A brief survey of radio access network backhaul evolution: part i. *IEEE Communications Magazine* 49 (6): 164–171. https://doi.org/10.1109/MCOM. 2011.5784002.

3 Osseiran, A. (2014). Mobile and wireless communications system for 2020 and beyond(5G).https://www.metis2020.com/wp-content/uploads/presentations/ITU-R-2020-VisionWS.pdf.

4 Andrews, J.G., Buzzi, S., Choi, W. et al. (2014). What will 5G be? *IEEE Journal on Selected Areas in Communications* 32 (6): 1065–1082.

5 Ahsan Kazmi, S.M., Tran, N.H., Ho, T.M. et al. (2016). Decentralized spectrum allocation in D2D underlying cellular networks. *2016 18th Asia-Pacific Network Operations and Management Symposium (APNOMS)*, pp. 1–6. IEEE.

6 Kazmi, S.M.A., Tran, N.H., Saad, W. et al. (2017). Mode selection and resource allocation in device-to-device communications: a matching game approach. *IEEE Transactions on Mobile Computing* 16 (11): 3126–3141.

7 Ullah, S., Thar, K., and Hong, C.S. (2017). Cache decision for scalable video streaming in information centric networks. *Multimedia Tools and Applications* 76 (20): 21519–21546.

8 Ahsan Kazmi, S.M., Tran, N.H., Ho, T.M. et al. (2015). Resource management in dense heterogeneous networks. *Network Operations and Management Symposium (APNOMS), 2015 17th Asia-Pacific*, pp. 440–443. IEEE.

9 Olwal, T.O., Djouani, K., and Kurien, A.M. (2016). A survey of resource management toward 5G radio access networks. *IEEE Communications Surveys Tutorials* 18 (3): 1656–1686. https://doi.org/10.1109/COMST.2016.2550765.

10 Rappaport, T.S., Xing, Y., MacCartney, G.R. et al. Overview of millimeter wave communications for fifth-generation (5G) wireless networkswith a focus on propagation models. *IEEE Transactions on Antennas and Propagation* 65 (12): 6213–6230. https://doi.org/10.1109/TAP.2017.2734243.

11 Ho, T.M., Tran, N.H., Ahsan Kazmi, S.M. et al. (2016). Distributed resource allocation for interference management and QoS guarantee in underlay cognitive

femtocell networks. *2016 18th Asia-Pacific Network Operations and Management Symposium (APNOMS)*, pp. 1–4. IEEE.

12 Ahsan Kazmi, S.M., Tran, N.H., Saad, W. et al. (2016). Optimized resource management in heterogeneous wireless networks. *IEEE Communications Letters* 20 (7): 1397–1400. https://doi.org/10.1109/LCOMM.2016.2527653.

13 Adedoyin, M.A. and Falowo, O.E. (2017). QoS-based radio resource management for 5G ultra-dense heterogeneous networks. *2017 European Conference on Networks and Communications (EuCNC)*, pp. 1–6. doi: 10.1109/EuCNC. 2017.7980721.

14 Visual Networking Index Cisco (2016). Global mobile data traffic forecast update, 2015–2020 White Paper. Document ID 958959758.

15 Li, L., Zhao, G., and Blum, R.S. (2018). A survey of caching techniques in cellular networks: research issues and challenges in content placement and delivery strategies. *IEEE Communications Surveys Tutorials* 20 (3): 1710–1732. https://doi. org/10.1109/COMST.2018.2820021.

16 Yang, C., Yao, Y., Chen, Z., and Xia, B. (2016). Analysis on cache-enabled wireless heterogeneous networks. *IEEE Transactions on Wireless Communications* 15 (1): 131–145. https://doi.org/10.1109/TWC.2015.2468220.

17 Ji, J., Zhu, K., Ran, W. et al. (2018). Energy efficient caching in backhaul-aware cellular networks with dynamic content popularity. *Wireless Communications and Mobile Computing* 2018: 1–12.

18 Satyanarayanan, M., Bahl, P., Caceres, R., and Davies, N. (2009). The case for VM-based cloudlets in mobile computing. *IEEE Pervasive Computing* 8 (4): 14–23. https://doi.org/10.1109/MPRV.2009.82.

19 Cisco (2015). Fog computing and the Internet of Things: Extend the cloud to where the things are. White Paper.

20 Beck, M.T., Werner, M., Feld, S. et al. (2014). Mobile edge computing: A taxonomy. *Proceedings of the Sixth International Conference on Advances in Future Internet*, pp. 48–55. Citeseer.

21 Mao, Y., Zhang, J., and Letaief, K.B. (2016). Dynamic computation offloading for mobile-edge computing with energy harvesting devices. *IEEE Journal on Selected Areas in Communications* 34 (12): 3590–3605. https://doi.org/10.1109/ JSAC.2016.2611964.

22 Messous, M., Sedjelmaci, H., Houari, N. et al. (2017). Computation offloading game for an UAV network in mobile edge computing. *2017 IEEE International Conference on Communications (ICC)*, pp. 1–6. doi: 10.1109/ICC.2017.7996483.

23 Kazmi, S.M.A., Tran, N.H., Ho, T.M., and Hong, C.S. (2018). Hierarchical matching game for service selection and resource purchasing in wireless network virtualization. *IEEE Communications Letters* 22 (1): 121–124.

24 Ahsan Kazmi, S.M. and Hong, C.S. (2017). A matching game approach for resource allocation in wireless network virtualization. *Proceedings of the 11th*

International Conference on Ubiquitous Information Management and Communication, p. 113. ACM.

25 Kim, D.H., Kazmi, S.M., and Hong, C.S. (2018). Cooperative slice allocation for virtualized wireless network: A matching game approach. *Proceedings of the 12th International Conference on Ubiquitous Information Management and Communication*, p. 94. ACM.

26 Sciancalepore, V., Zanzi, L., Costa-Perez, X. et al. (2018). ONETS: Online network slice broker from theory to practice. arXiv preprint arXiv:1801.03484.

27 Richart, M., Baliosian, J., Serrat, J., and Gorricho, J.-L. (2016). Resource slicing in virtual wireless networks: a survey. *IEEE Transactions on Network and Service Management* 13 (3): 462–476.

28 Foukas, X., Patounas, G., Elmokashfi, A., and Marina, M.K. (2017). Network slicing in 5G: survey and challenges. *IEEE Communications Magazine* 55 (5): 94–100.

29 Zhang, H., Liu, N., Chu, X. et al. (2017). Network slicing based 5G and future mobile networks: mobility, resource management, and challenges. *IEEE Communications Magazine* 55 (8): 138–145.

30 Ordonez-Lucena, J., Ameigeiras, P., Lopez, D. et al. (2017). Network slicing for 5G with SDN/NFV: concepts, architectures and challenges. arXiv preprint arXiv: 1703.04676.

31 Ho, T.M., Tran, N.H., Ahsan Kazmi, S.M. et al. (2018). Wireless network virtualization with non-orthogonal multiple access. *NOMS 2018–2018 IEEE/IFIP Network Operations and Management Symposium*, pp. 1–9. IEEE.

32 Vo, P.L., Nguyen, M.N.H., Le, T.A., and Tran, N.H. (2018). Slicing the edge: Resource allocation for RAN network slicing. *IEEE Wireless Communications Letters* 7 (6): 970–973.

33 Bastug, E., Bennis, M., Médard, M., and Debbah, M. (2017). Toward interconnected virtual reality: opportunities, challenges, and enablers. *IEEE Communications Magazine* 55 (6): 110–117.

34 Boyd, S., Parikh, N., Chu, E. et al. (2011). Distributed optimization and statistical learning via the alternating direction method of multipliers. *Foundations and Trends in Machine Learning* 3 (1): 1–122. https://doi.org/10.1561/2200000016.

Further Reading

Lin, T., Ma, S., and Zhang, S. (2015). On the global linear convergence of the ADMM with multiblock variables. *SIAM Journal on Optimization* 25 (3): 1478–1497.

Manzoor, A., Tran, N.H., Saad, W. et al. (2019). Ruin theory for dynamic spectrum allocation in LTE-U networks. *IEEE Communications Letters* 23 (2): 366–369.

Nishihara, R., Lessard, L., Recht, B. et al. (2015). A general analysis of the convergence of ADMM. arXiv preprint arXiv:1502.02009.

Shen, C., Chang, T.-H., Wang, K.-Y. et al. (2012). Distributed robust multicell coordinated beamforming with imperfect CSI: an ADMM approach. *IEEE Transactions on Signal Processing* 60 (6): 2988–3003.

Wang, C., Liang, C., Richard Yu, F. et al. (2017). Computation offloading and resource allocation in wireless cellular networks with mobile edge computing. *IEEE Transactions on Wireless Communications* 16 (8): 4924–4938.

Part IV

5G Verticals – Network Infrastructure Technologies

6

The Requirements and Architectural Advances to Support URLLC Verticals

Ulas C. Kozat[1], Amanda Xiang[2], Tony Saboorian[2], and John Kaippallimalil[2]

[1] *Futurewei Technologies, Santa Clara, CA, USA*
[2] *Futurewei Technologies, Plano, TX, USA*

Abstract

This chapter provides a self-contained overview of the vertical requirements as well as the technical and architectural concepts that is critical in delivering ultra-reliable and low-latency communications (URLLC) services. It presents a number of popular URLLC use cases. The chapter focuses on the new 5G network architecture and how it can be inter-faced with URLLC applications. It explains the evolution of software defined networking and network function virtualization concepts, 5G core network and notions such as application specific slicing. A cloud-native architecture divides the tightly integrated and virtualized network functions-specific software stack into more scalable, reusable and efficiently deployable software components. The implications of cloud native network functions and architecture on provisioning URLLC services should be carefully considered. The 5G system service-based architecture and extensions for vertical and URLLC applications are under discussion in 3GPP standards.

Keywords   *5G network architecture; cloud native architecture; deployable software components; software stack; ultra reliable and low latency communications vertical; virtualized network functions*

5G Verticals: Customizing Applications, Technologies and Deployment Techniques, First Edition.
Edited by Rath Vannithamby and Anthony C.K. Soong.
© 2020 John Wiley & Sons Ltd. Published 2020 by John Wiley & Sons Ltd.

6.1 Introduction

Vertical applications will be the main driving force for 5G deployments, and ultra-reliable and low-latency communications (URLLC) capability is one of the key 5G features attracting many vertical industries and applications, such as industrial Internet of Things (IoT), smart grid, vehicle to X, and professional multi-media production industry. Because of the nature of their application requirements, these vertical industries provide unprecedented use cases and requirements for 5G URLLC services, which guides the communication industry to develop suitable 5G technologies and deployment options. Our goal in this chapter is to provide a self-contained overview of the vertical requirements as well as the technical and architectural concepts that will be critical in delivering URLLC services.

The organization of this chapter is divided into two parts. In the first part of the chapter, a number of popular URLLC use cases are presented. As will become clearer later in that part, URLLC use cases are quite broad and even within the same vertical use case (e.g. factory automation) there are various payloads with quite diverse requirements. Furthermore, some of these vertical use cases have quite different privacy, isolation, and control requirements impacting their network deployment scenarios. The most recent focus in standardization organizations is overviewed accordingly.

In the absence of a single dominating killer application, a network architect can design a URLLC service either based on the most stringent requirements or design a system that covers as many use cases as possible under the cost budget. Neither of these options, however, is desirable as they do not reflect the evolving market needs appropriately. As standardization and deployment of networks that cover these cases take many years, once the network is deployed, it has to generate net revenue from the considered use cases for further investment, which can cripple the adoption of new services and creation of new markets on time. This brings us to the notion of decoupling the infrastructure from the network functionality and services. While the former is purely about the fixed assets that are once deployed, they must have a long enough lifetime to amortize their investment, the latter constitutes the real added value that the networks are providing. Thus, the notion of network programmability and more modular network stack through functional disaggregation, network virtualization, network function virtualization (NFV), and software defined networks are indispensable for the overall success of URLLC services.

In the second part of the chapter, the focus is therefore shifted toward the new 5G network architecture and how it can be interfaced with URLLC applications. The evolution of software defined networking SDN and NFV concepts, 5G core network and notions such as application specific slicing are explained in more detail. In particular, the content covers: (i) the three phases of SDN research and development, (ii) NFV background and its evolution into the cloud native era, (iii) the functional disaggregation, flattened and micro-service-based evolution of the control plane, control plane and user plane separation, and mobility in 5G core networks, and (iv) network slicing as the key interfacing between URLLC applications and the network.

By the end of the chapter, the reader will gain significant insights on URLLC use cases and the key architectural concepts needed for their success.

6.2 URLLC Verticals

In this section, we summarize a few of the new use cases proposed by the vertical industry players that drive the 5G URLLC discussions in standardization organizations and industrial forums.

6.2.1 URLLC for Motion Control of Industry 4.0

Industry 4.0, or smart manufacturing, is becoming a global industry trend to transform traditional factories into smart factories. Instead of being fixed constructs, these smart factories have modular structures and can be reorganized according to evolving business and manufacturing processes. They are augmented to cyber-physical systems that monitor physical processes, create a virtual copy of the physical world, and make decentralized decisions. Following the IoT paradigm, cyber-physical systems communicate and cooperate with each other and with humans in real-time. Such communication and cooperation occurs both internally and across organizational boundaries. Services are offered and used by participants of the value chain [1]. As a mandatory step toward such a transformation, smart factories must host highly reliable and fast communication networks. Some use cases in Industry 4.0, such as motion control in the assembly line, provides very rigid performance targets for 5G URLLC services as explained next.

A motion control system is responsible for controlling the moving and/or rotating parts of machines in a well-defined manner, for example in printing machines, machine tools, or packaging machines. In a typical assembly line, the motion control system contains a closed loop control system with controller, actuator, and sensors (Figure 6.1) [2].

The interaction among components follows a strictly cyclic and deterministic pattern with a short cycle time (it can be as short as 0.5 ms), and these interactions rely

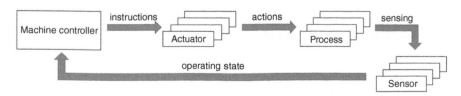

Figure 6.1 Motion control system in factory assembly line.

on a URLLC link. Table 6.1 presents three application scenarios and communication service requirements. Considering that these applications require ultra-low latency for transmitting the control information between the components and their deployments are local and close to each other within the boundaries of the service area, an edge cloud collocated with the access network functions is naturally desired.

6.2.2 Multi-Media Productions Industry

Professional culture and creative multi-media industry has been using wireless technologies for a while, such as using wireless for audio (e.g. microphones, in-ear monitor), video (e.g. cameras, displays, and projectors), and stage control systems for concerts, filming, and conferences. However, the existing wireless, such as WiFi, Bluetooth, and 4G technologies have their limitations, especially the lack of high reliability and low-latency capabilities. Therefore, the industry is looking into 5G to provide the industry the suitable communication services that can satisfy the demanding requirements of these professional applications. Let us examine one such application, namely live performance. Figure 6.2 shows the system topology of a live audio performance [2].

In live performance scenarios, artists on stage carry wireless microphones while listening to their own voice through a wireless in-ear-monitor (IEM) that is connected to the live audio mixing machine (the artist's voice is fed back to the mixer and transferred back to the artist's ear). Artists may be moving with considerable speeds (up to 50 km/h) around the service area, for instance when they wear roller skates. In the live performance case, the latency of the wireless audio-production system is much lower and the reliability requirement much higher than other speech transmissions systems because not only does live performance require higher quality audio and real-time performance to give the best experience to the audience but also the audio source (e.g. a microphone) and the audio sink (e.g. an IEM) are co-located on the head of the performing artist. Due to human physiology (i.e. cranial bone conduction), an end-to-end application latency greater than 4 ms would confuse the performing artist while hearing through the IEM.

Table 6.2 summarizes some of the typical performance requirements for the live performance application [3]. Note that the wireless link latency here is naturally much stricter than the user experienced mouth-to-ear latency budget of 4 ms that includes application processing and interfacing latencies.

6.2.3 Remote Control and Maintenance for URLLC

As more augmented reality (AR) devices, robots, and automation systems are being deployed in smart factories and power plants, conducting the operations and maintenance remotely from a control center becomes more desirable to the industry from the cost, safety, and productivity perspectives. For example, the engineers can

Table 6.1 Examples of motion control scenarios and characteristics (3GPP TR 22.804) [2].

Application	No. of sensors / actuators	Typical message size (byte)	Transfer cycle time T_{cycle} (ms)	Service area	End-to-end latency: target value	Communication service availability (%)
Printing machine	>100	20	<2	100 m × 100 m × 30 m	<Transfer interval	999 999–99 999 999
Machine tool	~20	50	<0.5	15 m × 15 m × 3 m	<Transfer interval	999 999–99 999 999
Packaging machine	~50	40	<1	10 m × 5 m × 3 m	<Transfer interval	999 999–99 999 999

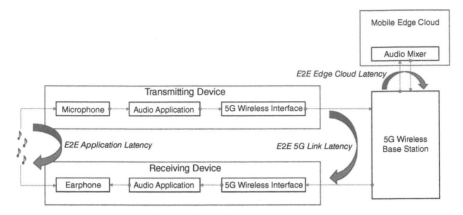

Figure 6.2 System diagram for live audio performance.

Table 6.2 Quality of service requirements for live performance.

	Characteristic system parameter
End-to-end latency	<600 μs
Audio data rate	100 kbit/s to 54 Mbit/s
Packet error ratio	$<10^{-4}$
No. of audio links	5–300
Service area	$\leq 10\,000\,\text{m}^2$
User speed	$\leq 50\,\text{km/h}$

reside in the central office using AR devices to stream video from a remote and hazardous plant (e.g. mining or drilling site, chemicals factory, etc.) while controlling one or more robots at the same time to conduct maintenance and repair work (e.g. fixing leaks, cooling systems, ventilation, etc.). In some deployments, the distance between the remote plant and the central control office can be up to 5000 km [2]. In this case, highly reliable (999 999–999 999 999% service availability), low latency (end-to-end latency <100 ms) and long distance communication links (up to 5000 km) that may span different regions or wide area network are desired.

6.2.4 Vehicle-to-Everything

Vehicle-to-everything (V2X) encompasses various type of vehicular application namely vehicle-to-vehicle (V2V), vehicle-to-infrastructure (V2I), vehicle-to-network (V2N), and vehicle-to-pedestrian (V2P). As the industry moving closer to the premise of the connected and autonomous vehicles, the V2X applications are evolving and the performance requirements are becoming more constrained.

Basic requirements for exchange of status information of vehicles, position, speed, direction of travel, etc., are strengthened by higher reliability and lower latency targets to meet advanced and emerging V2X applications. The requirements are becoming more restricted with the levels of automation (LoA). The SAE International Standard J3016 defines six levels of driving automation with performance requirements ranging from a human driver to a fully automated system. The LoA are: 0, no automation; 1, driver assistance; 2, partial automation; 3, conditional automation; 4, high automation; and 5, full automation.

3GPP has defined the cellular standards for LTE-based V2X communication and is currently working on defining 5G-based V2X communication to meet the advanced requirements. Vehicles Platooning, Advanced Driving (i.e. semi-automated or fully automated driving), Extended Sensors, and Remote Driving are among V2X scenarios that benefit from the 5G technology. As an example, 3GPP TS22.186 (Enhancement of 3GPP support for V2X scenarios – Stage 1, table 5.5-1) defines the performance requirements for remote driving as maximum end-to-end latency of 5 ms at a reliability of 99.999% and at speeds up to 250 km/h, for information exchange between a user equipment (UE) supporting V2X application and a V2X Application Server.

6.3 Network Deployment Options for Verticals

Traditionally, public mobile network operators (PMNOs) provide the network communication service to the vertical industry companies using cellular technologies to meet their connectivity needs, or the vertical players to develop and manage their own network using unlicensed spectrum, such as WiFi. The existing deployments, however, are rigid and cannot meet the fast growing communication service needs from the verticals, especially to meet the custom and vast requirements on performance, security and privacy dictated by the new business models of the verticals. There is a strong emerging demand from vertical players that there should be a dedicated and isolated URLCC 5G network for them to run their specialized applications. 5G not only introduces new disruptive technologies, but also provides flexible deployment options for new networks, e.g. verticals can own and manage their 5G URLCC network over their own or leased spectrum, or run their applications over specific network slices provisioned by PMNOs. Currently, there are several potential network deployments options for verticals being studied and developed in the industry, which can be divided into two main categories:

1) PMNOs work with vertical service owners to provide the desired network services for the verticals.
2) Vertical service owners run their isolated and dedicated private 5G URLCC networks without any involvement of the PMNOs.

In the first category, PMNOs would work with verticals to provide certain communication infrastructure and communication service based on the service agreement. Because the data privacy, security, and service assurance for some high priority traffic are critical to the verticals, certain isolation between the vertical's traffic with the rest of the traffic running on the same public mobile network is required. Mobile network operators and verticals have the following deployment options where Multi-Operator Core Network (MOCN) [4] and network slicing [5–7] have been standardized but need to be further enhanced in order to meet the new requirements from verticals:

1) Mobile operator creates Virtual Private Network (VPN) on top of the underlying public network for the verticals by using the well-defined APN (Access Point Name) concept, where the traffic of the vertical service owner would be associated with and routed within the mobile operator's public network to the vertical's application server. In this option, verticals do not need their own mobile radio access network (RAN) or core network (CN). Instead, they reuse the network of the mobile operator.
2) MOCN where the vertical would share the same RAN and same spectrum owned and operated by mobile operators but verticals would have their own mobile CN. The vertical's traffic will be separated and routed from the shared RAN to the vertical's own mobile CN.
3) MORAN (Multi-Operator RAN) is similar to the MOCN case but the main difference is that (in MORAN case) the vertical would have a dedicated frequency in the shared RAN. Thus, MORAN provides better isolation and service assurance for vertical traffic than MOCN.
4) Mobile network operator uses network slicing technology to create a dedicated network slice for the vertical network on top of its public network infrastructure.

In the second category, verticals would (i) acquire a dedicated spectrum, (ii) lease the spectrum from the mobile network operator, or (iii) rely on third party neutral host operator to construct and manage its own private network. In some of these private network deployments, verticals may deploy their own non-3GPP compliant authentication and authorization mechanisms to leverage their existing non-cellular networking stack. This category of private network deployment options is considered as completely isolated from PMNOs, and may or may not have interaction or service continuity with the public mobile networks. There are also industry efforts being initiated in several industry consortiums and standard organizations to identify and develop the potential solutions to achieve service continuity and mobility between certain private network deployment and public mobile network, such as 5G ACIA, 3GPP, CBRS, and MultiFire.

As covered so far, URLLC vertical has quite broad application requirements and deployment scenarios. Combined with the other services such as enhanced mobile broadband (eMBB) and massive IoT, 5G networks should be very flexible in terms of how services are defined and deployed rather than providing a single network solution for all services. As SDN and NFV concepts have been the key technology areas to bring such flexibility into the communication networks, the next section takes a closer look at SDN and NFV topics as well as how they are adopted in 5G architecture in the context of supporting URLLC verticals.

6.4 SDN, NFV and 5G Core for URLLC

It is widely acknowledged that the 5G system is more than a new radio interface, it is about the new network infrastructure, where programmability, virtualization, network slicing, and cloudification are widely acknowledged and intertwined themes. This section is divided into three parts that overview SDN, NFV, and 5G core concepts to support URLLC services.

6.4.1 SDN for URLLC

SDN simplifies the transport network by removing intelligence from the individual forwarding elements to logically centralized network controllers through which all the control plane functions interact with the state of the forwarding elements. Such an architecture allows to run network controllers and control plane functions over commodity servers and add more services into the control plane in a much more scalable manner independent of the forwarding plane. Furthermore, forwarding plane capacity can be expanded independently as more bandwith is needed.

The first generation of SDN solutions based on OpenFlow specification required switch vendors to follow a particular abstraction: first a single flow table then a pipeline of flow tables and metadata that can be programmed with match and action rules on network flows. First OpenFlow specifications fixed the set of protocols and fields that can be parsed, modified, added, and removed. Later specifications allowed for extensibility of north bound interface (NBI) as more protocols and header fields are supported by vendors.

The second generation of SDN solutions based on OpenFlow or NetConf interfaces allowed modeling of internal pipeline and programmability (e.g. APIs) of each forwarding element separately rather than fixing these models as in the case of the first generation. Thus, network controllers could control a heterogeneous set of network equipment varying from physical layer (e.g. optical and microwave equipment), layer 2 (e.g. Ethernet, MPLS), layer 3 (e.g. IPv4, IPv6), and layer 4 (e.g. TCP, UDP).

The third generation of SDN solutions brought more protocol independence by making the pipeline of the forwarding element itself programmable while deployed on the field (e.g. protocol-oblivious switching [8], P4 [9], OpenState [10], Programmable Buffers [11], etc.). Thus, it has become possible to support new user plane functionality on the transport network without changing any deployed physical equipment.

The first two generations of SDN solutions greatly helped toward the management and orchestration of network services as it made it simpler to program/change the forwarding behavior of the network while making it possible to run computation and data intensive operations outside the CPU and memory constrained forwarding elements. As SDN flattened the forwarding topology, it opened up new avenues for providing flexibility in traffic engineering and steering in the forms of multi-path routing, load balancing, anycasting, network slicing, etc., to support network flows that require URLLC services. The third generation of SDN solutions further enabled on path and in-band solutions to the problem where the control plane itself could be the bottleneck in reacting to network events such as link failures/degradations or sudden changes in traffic patterns. Thus, URLLC services can be better supported.

SDN provides a unified picture to program and configure the underlay (i.e. the transport fabric) and overlay network functions (e.g. all the functions running over the IP network). NFV as covered in the next section is considered as a way of evolving the overlay functions, while SDN is considered as the piece that stitches these overlay functions together and configures them to deliver a particular network service.

6.4.2 NFV for URLLC

6.4.2.1 NFV Background
As the wireless access moves toward a cloud-based infrastructure, many network functions are no longer hardwired to specific physical nodes in the network. Instead, they become Virtualized Network Functions (VNFs) that can run anywhere in the wireless access cloud and service capacity can be scaled based on the actual realized demand. This flexibility, however, comes with a performance penalty. The first iteration for NFV has heavily relied on virtual machines (VMs) as many network functions are designed as software monoliths that work on a particular hardware configuration and with a particular software stack. The standard ETSI NFV models network services in the form of a topology graph with VNFs as the vertices of that service graph [12]. Furthermore, VNFs themselves are modeled as a topology graph with VNF components (VNFCs) corresponding to the vertices in that graph. Figure 6.3 depicts the high-level model for a network service example with three VNFs. Each VNF has external and internal (if modeled with more

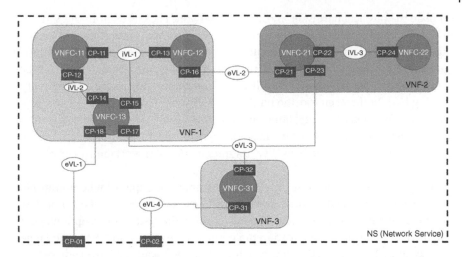

Figure 6.3 ETSI model for network service and virtualized network functions.

than one VNFC) virtual links, which typically corresponds to a L1, L2 or L3 network or a point-to-point connection. In the figure, these external and internal virtual links are enumerated with prefixes eVL and iVL, respectively. If the underlying networking infrastructure is based on SDN, such virtual links (as specified in network service and VNF models) are created and managed by network controller(s). Each VNFC instance is encapsulated in a software container (e.g. VM image) called a Virtualization Deployment Unit (VDU). In many cases (e.g. firewall, customer premise equipment), a VNF is simply composed of one VNFC. In mobile networks, they may have more complex structures with many VNFCs (e.g. vEPC, vIMS, vRAN, etc.). How network functions are bundled together or disaggregated can be vendor or operator specific. The ETSI model is generic enough to cover both legacy and future deployment needs. Although not shown in the figure, ETSI model also incorporates VNF forwarding graphs, where different service function chains can be defined for distinct groups of network flows. Independent of VNF complexity, there is a one-to-one relationship between a VNFC instance and a VM instance. As such, the choice of virtualization technologies and stack have direct impact on network latency, throughput and isolation.

6.4.2.2 Reducing Virtualization Overhead on a Single Physical Server

As VNFs are deployed on commodity servers, additional processing latency is incurred due to the virtualization layer. Figure 6.4 provides a high-level view of various external and internal networking scenarios. In Figure 6.4a, the VM is connected to the external networking fabric through the physical hardware (i.e. network interface card). For maximum portability, the virtualization layer can emulate the physical hardware. In this case, network packets to/from the VM follow path 1 in the figure. This path requires CPU interrupts, multiple memory copying, and context switching. Utilizing mature technologies such as SR-IOV, compatible hardware and modified drivers in VMs, the virtualization layer can be completely bypassed (path 2) preventing CPU interrupts and unnecessary memory copying [13]. In this scenario, the hardware itself supports virtualization and provides per VM packet filtering, buffering, and forwarding. Since the virtualization layer is bypassed and hardware dependency is formed, important features of machine virtualization such as live migration and VM portability cannot be supported.

In Figure 6.4b, a service function chaining scenario is depicted where multiple functions are collocated on the same physical host and process packets of a flow in a chain. When a packet is passed from one VM to the next, various options exist and three of them are enumerated in the figure. In the first option, VMs exchange packets directly via shared memory without any software or hardware switch involvement [14]. This is the fastest option as it eliminates unnecessary memory copies across stacks and interrupts. However, memory isolation is violated and control/management of function chaining becomes more convoluted. In the second option, VMs exchange packets through the virtualization layer utilizing a software switch. This is a relatively slower option particularly for small payloads (e.g. 64 bytes) and longer chains (e.g. one to two orders of magnitude less throughput) even when the software switch is implemented using technologies such as DPDK. Hence, the processing overheads and extra memory copying are still substantial in this option [14, 15]. Nonetheless, it provides a clear separation, extensibility, and controllability. In the third option, VMs exchange packets through the physical

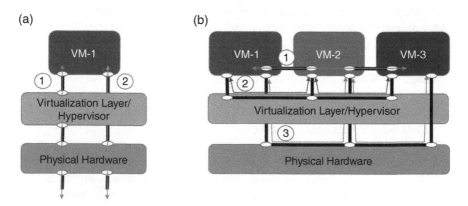

Figure 6.4 Various external and internal networking scenarios for VMs exist with different latency, throughput and isolation properties. (a) External networking and (b) function chaining on the same host.

hardware bypassing the networking stack in the virtualization layer. This option is fast when packets are outbound/inbound from/to the physical host and few functions in the same chain are collocated. For longer chains, however, this option reaches the limits of the hardware capabilities and does not scale well.

Network functions need to process millions of packets per second to support data speeds in the order of Gbps. To achieve such fast data paths, there are a number of technical concepts that are applied by different solutions [1, 13–18] and we summarize them in the following paragraphs.

Batching. Interrupting CPU processing or performing context switching at such high rates would starve the network functions to perform their tasks. Aggregating processed packets in output buffers at each layer and function, then passing them as a batch to the next layer and function for batch processing substantially reduces the overhead of interrupts and context switching. Vector Packet Processing (VPP) for instance deploys such batch processing aggressively [1]. Batching can be done based on a mixed policy of maximum batch size and maximum timeout value to avoid unnecessary latency. When traffic rate is high, the maximum batch size limits the packet latency. When traffic rate is low (thus it takes a long time to reach the maximum batch size), maximum timeout value limits the packet latency.

Polling. Instead of resorting to I/O interrupts to notify the next network function about the presence of a new packet (or batch of packets) to process, network traffic can be buffered until the next function is scheduled and it polls for the packets [15]. Polling can be done dynamically. When traffic rate is high, every scheduled function will likely have a backlog of traffic and thus the system can achieve maximum throughput. In contrast when traffic rate is low, the CPU resources are heavily under-utilized, thus rather than waiting for the network function to poll for the packets, an I/O interrupt would reduce the packet latency.

Zero-copy. Every time a packet is copied from one memory location to another, it adds significant overhead without actually performing useful work on the packet. If the NFV stack is not optimized, a packet can be easily copied several times, e.g. between physical network interface controller (NIC) and host machine kernel, between host machine kernel and guest machine virtual NIC, between virtual NIC and guest machine kernel, between guest machine kernel and guest machine user space, etc. At the expense of additional complexity, zero-copy-based systems do not copy the packets from one layer to the next or from one function to the next but provide libraries for direct memory access for reading, modifying and redirecting packet contents [15, 18].

Run-to-completion. When user plane functions (UPFs) perform small and well-defined tasks (e.g. checking IP header and selecting output port), they require a small and well-defined number of computational cycles. Under such circumstances, it is not worth pre-empting a currently executing function in order to schedule another function with a newly arriving but higher priority packet. The savings for the higher priority packet would be negligible, thus the tasks queued for pre-empted function would experience unwarranted delay spurs. Therefore, to streamline the packet processing and provide a steady progress, it is preferable to avoid the overheads of context switching and pre-emption. Many solutions such as DPDK, NetBricks, and SoftNIC adopt this method [15–17].

Hardware offloading. All the techniques summarized so far fundamentally try to eliminate the idle processing times and the processing cycles wasted toward non-application related tasks. Even when all such waste is prevented, a software-based solution over a general-purpose CPU cannot match the speed, density, and power efficiency of a purpose-built hardware-based solution. Offloading tasks such as encryption/decryption, check sum computation, compression, etc., are typical examples for computationally heavy and repeatedly performed tasks in the data plane. Programmable hardware such as P4, FPGA, and GPUs are also getting more attention for deploying hardware accelerated network functions [9, 19, 20].

6.4.2.3 Evolution of NFV toward Cloud Native Network Functions

Decoupling network functions from a specific hardware instance is not the end target but a critical step toward converting access and CNs into a utility-based computing, storage and communication infrastructure. Rather than dividing the network into silos of integrated boxes, in a cloudified architecture, all the underlying hardware resources are pooled together and utilized as needed, where needed, and in a fine-grain fashion by many network services. A cloud-native architecture further divides the tightly integrated and VNF-specific software stack into more scalable, reusable, and efficiently deployable software components. Adding and removing network function instances without disrupting existing services and user sessions becomes the new norm of scaling the service capacity. Seamless load balancing, routing, traffic engineering, and traffic migration should be supported by the network architecture in support of such capacity scaling. In return, cloud-native network functions should handle dynamics of instances being popped up or destroyed without any side effects on other instances and ongoing sessions.

Figure 6.5 shows an example service model for cloud native architecture that is being utilized in popular cloud platforms such as Kubernetes [21]. The ETSI NFV model supports a broad class of network services and transport technologies. Thus, operators can utilize their existing transport networks that can be quite heterogeneous. In contrast to the ETSI NFV model for network services and functions depicted in Figure 6.3, the service model in Figure 6.5 provides a much simpler and cleaner service infrastructure as it assumes a homogeneous transport network (e.g. IPv4 or IPv6) and no-state sharing across deployment instances. The overall emphasis is on how to group functional components, how to scale them, and how to hide cardinality of each functional group behind a single connection used as a service access point. Functional components that are grouped together

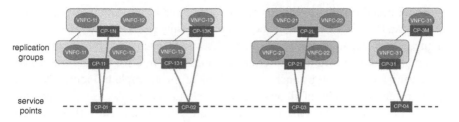

Figure 6.5 An example service model for cloud native architecture.

form a replication group (or unit) that is scaled in that granularity (e.g. leftmost replication group has VNFC-11 and VNFC-12, thus adding one more unit adds one more VNFC-11 and VNFC-12). Each functional group shares the same memory space and network name space, allowing very fast information exchange/ sharing across components in the same group. Each group has a single connection point to the external network and that connection is also shared across the functional components in each group. Cloud native platform distributes incoming traffic on service connection points (e.g. CP-01 through CP-04) toward the actual VNFC groups (e.g. CP-11 through CP-1N for the first replication group). State-of-the-art platforms perform a load balancing service based on various scheduling policies that enable control over what percentage of traffic is steered toward each group. Such control enables not only more even load distribution but it also simplifies the process of phasing in the new software and phasing out the old software for the VNFC groups.

The implications of *cloud native* network functions and architecture on provisioning URLLC services should be carefully considered. On the one hand, cloud native systems make it simpler to launch many function instances over a distinct set of physical resources dynamically as a function of load, proximity, and efficiency. As such, it can provide a higher resilience against hardware failures and it also makes it easier to bypass hot spot areas for URLLC services. Figure 6.6 depicts the situation where three VNFCs that have high inter-component network traffic utilize the same amount of computing resources in a non-cloud native system (upper row) and cloud native system (lower row). A cloud native system can sustain network service up to two host failures, whereas a non-cloud native system cannot handle any host failure. Furthermore, a cloud native system can pack component instances that have high traffic volume among each other on the same host, hence facilitating much higher throughput and reduced latency. On the other hand, cloud native systems may have performance issues in this particular set up as follows: Even when network functions are stateless, as soon as some VNFCs are bundled together to optimize the networking performance that would create a state for session processing in the system such that a flow packet processed by a first function on a given host must be processed by a second function on the same host. Otherwise, bundling does not provide any gains in networking performance. Thus, when we admit a number of sessions, each session is assigned to a chain and cannot move to another chain unless it is interrupted and redirected. Assuming an even load balancing of sessions in the example in Figure 6.6, each chain receives one third of all sessions in the cloud native case versus in the non-cloud native case all sessions are served by the same chain. Thus, there is more statistical multiplexing in the case where VNFCs are not bundled. Nonetheless, this is not a shortcoming of the cloud native design but of the particular configuration and placement choice. Alternatively, all VNFCs of the same type can be collocated to match the configuration of non-cloud native VNF deployments.

3GPP has already adopted SDN and NFV concepts along with micro-services in its specifications particularly for the CN as discussed next.

6.4.3 5G Core and Support for URLLC

3GPP specifications [5, 6] include control plane and user plane functionality that provide many basic capabilities to support low-latency and high-reliability application flows. Control plane functions in the mobile core have been

VS.

Figure 6.6 Cloud native systems can provide higher service resiliency and faster networking between VNFCs.

further disaggregated in comparison with 4G. To support low-latency and highly reliable applications, 3GPP has initiated new studies on key aspects for URLLC. Enhanced service-based architecture (eSBA) will study flexibility of UPFs and architectural support for micro-services and NFV in highly reliable deployments. Several other areas under study for URLLC include support for redundant transmissions, low latency and jitter during handover, session continuity during UE mobility, and QoS monitoring. SDN and NFV make it possible to address these issues in a flexible manner but also require the application of various techniques and optimizations to offer low-latency and high-reliability services.

Section 6.2 covered various use cases and requirements including motion control for Industry 4.0 and V2X. These services require short end-to-end user plane path to support high levels of QoS in terms of latency, bandwidth, and reliability. The control plane that manages the user plane for these services is also distributed significantly to be able to setup connections and manage high speed mobility.

3GPP TS 23.501 [5] specifies the architecture for the 5G system in detail. Figure 6.7 represents a subset of the functionality that is most relevant in the context of vertical URLLC services. This may represent a set of slices for URLLC that has specific supported features and network function optimizations as identified by S-NSSAI (Subscribed-Network Slice Selection Assistance Identifier) in the NSSF (Network Slice Selection Function). User plane flows between UE and application function (AF) (e.g. V2X) as well as among UEs (e.g. controller–sensor) need optimization since traversing the default IP gateway (UPF1 that serves as the session anchor point) would result in triangular routing. Low latency flows may

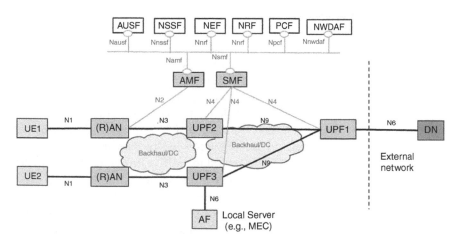

Figure 6.7 5G architecture.

also exist between two operators where the source UE and terminating UE are in separate service provider networks. Several other functions – Authentication Server Function (AUSF) for user authentication, NSSF for slice selection, Network Exposure Function for exposing capabilities to the application, Network Repository Function (NRF) for discovering function instances, Policy Control Function (PCF) for providing user and network policies, Network Data Analytics Function (NWDAF) for getting operator specific and slice specific data analytics – provide supporting services and are described further in TS 23.501 [5].

Many vertical applications require a short, low latency, QoS and policy constrained end-to-end transport path across radio and IP networks. The Access and Mobility Management Function (AMF) that is responsible for authorizing access and managing mobility of the UE should be able to cope with frequent and high rates of mobility for these applications. User plane path in the radio network (UE–RAN) is handled by the radio access controller and the SMF (Session Management Function) sets up the path across the UPF instances. Service end points are usually close to the UE – for example an end-to-end path from UE1 to UE2, or UE1 to AF. Disaggregation of control and user plane functions, and the ability to orchestrate them dynamically with NFV and SDN allow many possibilities for optimization.

SMF and UPF for a flow can be placed optimally. However, programming the forwarding plane and installing routes through the network for large numbers of flows that are also highly mobile is challenging. 5G specifications support three service and session continuity (SSC) modes: in mode 1, a PDU session anchor is kept for the duration of the session, in mode 2 (make-after-break) the network may signal the release of the PDU session and request UE to establish a new PDU session to the data network immediately after, and in mode 3 (make-before-break), users are allowed to have a new PDN session to the same data network before the old PDU session is released.

SSC modes 2 and 3 are useful for URLLC services that have to also support end-user mobility. Connectionless services or connection oriented services that are able to tolerate service interruption may use SSC mode 2. Services that are not tolerant of service interruptions can use SSC mode 3 and multi-homed transports such as MPTCP (Multipath TCP). For low latency end-to-end transport of application packets it is necessary to select the shortest paths that traffic engineered to support QoS, other policy constraints and which can be re-programmed as the UE moves. For the SMF to reprogram the UPF mid-session, they need to be flexibly provisioned, and NFV and SDN are essential tools. The flexibility that NFV and SDN provide are also necessary to keep the state associated with a session in close proximity. For example, in the flow in Figure 6.8, the (R)AN, AMF, SMF and multiple UPF instances contain soft-state for the PDU session. These disaggregated functions can be orchestrated to be proximate (i.e. it is a flexible composition of the functions) to deliver optimal control and user plane path.

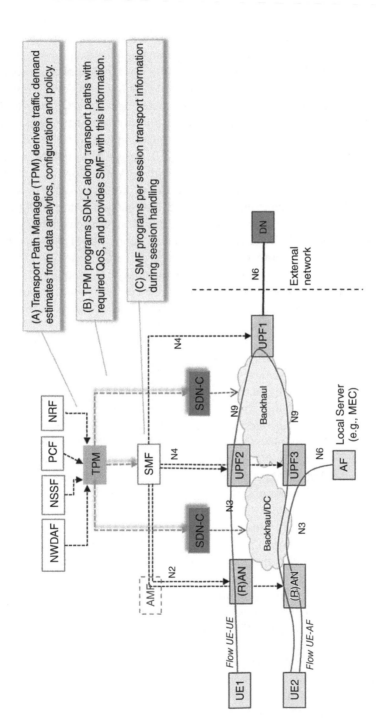

Figure 6.8 Transport path management.

Many end-to-end user plane paths for vertical applications terminate within the local network as these applications carry large amounts of data, or are highly latency sensitive. For low latency and high reliability, the transport paths need to be traffic engineered. Figure 6.8 shows a Transport Path Manager (TPM) function that derives traffic demand and programs the routers (via SDN-C) prior to the SMF setting up session transport paths during PDN attachment. Paths using default routes may not be the shortest – for example, in Figure 6.8, the default path between UE1 and UE2 are via UPF1 (PDU Session Anchor for both UEs). Flexibility in locating user plane functionality with NFV and SDN closer to the RAN can allow for more optimal paths. However, IP addresses allocated by highly distributed UPFs may change quickly as the UE moves to another UPF (more proximate but different IP subnet). For low latency services with high rates of mobility, multi-homed PDN connections and transports like MPTCP can be a better approach than handovers with IP address change. In all cases, IP transport congestion control actions due to slow-start, retransmission timeouts, and out-of-order delivery and jitter need to be considered. The control plane for managing mobility with IP address change requires a large amount of signaling during PDU session mobility. It is well known that the control plane operates at speeds that are orders of magnitude slower than the data plane [6]. NFV/SDN function placement as well as signaling mobility events in-band with the data packet can improve high-speed mobility performance.

The route management mechanisms for mobility considered so far have been based on anchored mobility. Anchor-based mobility mechanisms have been proven to scale well but they also require a large amount of signaling that can be a problem for high-speed mobility. With SDN and dynamic reprogramming of network functions, it is feasible to consider rendezvous or anchor-less mobility [22] or some combination to get better reactivity to fast mobility. However, there are many challenges in terms of scaling as well as modification to end-hosts (UE) that need further work.

End-to-end transport paths that provide guarantees in terms of QoS and resilience are needed in addition to route management. As a result of disaggregation and distribution of UPFs, the available backhaul transport path bandwidth between the micro-data centers can vary significantly even if the aggregate demand stays the same. Carrying traffic for content delivery, fixed and enterprise networks in an area on the same backhaul network allows for gains with statistical multiplexing. However, it is still necessary to dynamically provision transport paths that can meet the demand in terms of bandwidth and latency. This is illustrated in Figure 6.8 with the SDN Controller (SDN-C) that can provision requests in the backhaul and data center transports. The TPM collects input on expected session requirements from 3GPP functions such as SMF or NWDAF and provides this to the SDN-C to provision the traffic matrix. The TPM is a logical function that may be part of the NSSMF (Network Slice Selection Management Function). While the estimation of traffic demand and provisioning of resources is dynamic, i.e. in the order of several minutes, transient change in conditions of a path is signaled via IGP. IGP extensions to protocols such as OSPF to signal preferred path routing to satisfy a policy is being considered in the IETF routing area. These extensions along with the ability to signal OAM and route information in-band with the data packet allows for disaggregated functions like the UPF to implement policy-based routing at line speed.

The 5G control plane that is also disaggregated benefits from using NFV and SDN to orchestrate and place functions. For vertical applications, as with the user plane, the control plane is distributed such that the user plane can be controlled efficiently for users requiring low latency and high speed mobility. The 5G system SBA (service-based architecture) and extensions for vertical and URLLC applications are under discussion in 3GPP standards. It is possible to dynamically instantiate an NF instance and to scale the processing capacity of a function by adding or removing instances. Discovery of these function instances are managed by the NRF. The NRF maintains instance identifier, PLMN identifier, slice information, NF capacity, and IP address among other parameters. Since in an SBA these instances are dynamic, lookups and entry caching need to be enhanced. On path service routers that steer control plane signaling requests to the right instance can be much faster than a service discovery query-response followed by the signaling message. Performance with caching such dynamic instance information throughout the network versus using service routers and an intermediate hop for signaling requests need to be studied further.

Caching and replication of state information along with function virtualization can be used to design scalable and robust services. For the 5G system, subscriber information, policy, and other data that change less frequently can be cached – with cache coherency mechanisms to ensure that the latest copy of data is read. The volume of data for subscriber and policy information is relatively low and thus the storage overhead for multiple replicas should be acceptable. Signaling and memory/processing overheads are also manageable for synchronizing data that change less frequently (e.g. the system can be optimized for read heavy patterns by keeping write quorums large while read quorums small). This allows such functions to be distributed close to the control functions such as AMF or SMF that requests the data.

For data that changes during the handling of a session (mobility, connection data), it can be replicated to provide fault tolerance against failure of the server. In appliance-based realizations, as was common with 4G, fault tolerance is handled by replication across one or more cards within the system ($N+k$, hot or cold standby, etc.). The node (processing cards, system interconnect, etc.) is engineered to tolerate failures needed for the system. Network functions that are virtualized

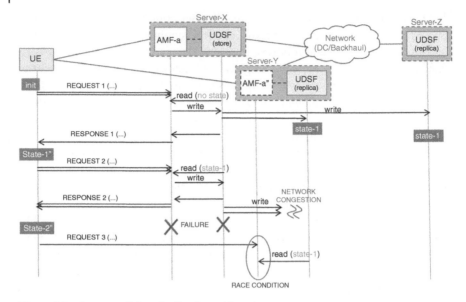

Figure 6.9 Race conditions in distributed functions.

on the other hand can replicate and store data across the network and be more resilient to single point failures. However, they must also be able to handle race conditions as a result of replicating state across a shared network. A simple example is shown in Figure 6.9.

Figure 6.9 is a simplified diagram that shows the outlines of a signaling exchange between the UE and AMF (to signal some access or mobility conditions). The AMF-a may contact other servers in the system before writing some new state and responding – these details are ignored in the diagram for simplicity. The AMF-a data stored in the UDSF on Server-X and replicated to UDSF in Server-Y and Server-Z with high system availability. Normally, AMF-a has session data to process requests from the UE (request/response 1–2). However, after AMF-a responds to the second sequence (RESPONSE 2), network congestion delays the replication of this information and Server-X fails. The NRF routes the next request to AMF-a″ on Server-Y which has a stale state (state-1) with which the subsequent request from the UE is processed.

In the above example, it may be noted that Server-X and Server-Y may be in the same data center and thus the chance of network congestion or failure there versus that across an access network maybe relatively rare. However, with increasing access speeds and design for vertical applications, these inherent conditions should be designed to be detected and handled in such a way that the processing is consistent.

As with control plane function virtualization, application services (AF in Figure 6.7) can be virtualized with caching and replication for scale and fault tolerance. The 5G system describes using NEF (Network Exposure Function) to expose network capability and events to the AF to support connection and mobility awareness. In this case, the capability is exposed by the control plane entities via a publish-subscribe mechanism. The subscription and publishing of information is handled on a per UE basis. An alternative approach to selecting AF would be by using IP anycast addressing. Depending on the location at which the UE requests the service, the IP routing can direct it to the appropriate/close instance of that AF. In case of failure (or mobility), a new instance of the AF is selected (e.g. via utilizing IGP routing updates and metrics). The IP anycast solution is known to be highly robust to failure and distributed denial-of-service attacks while being scalable as the updates happen per server instance and not per UE. The NEF approach on the other hand provides very fine grained control.

How 5G networks evolve is largely dependent on the successful interfacing between the applications and the network. We present a particular view on such interfacing in the next section.

6.5 Application and Network Interfacing Via Network Slicing

In mobile Internet, applications and networks are decoupled from each other; applications have no direct control over how their packets are processed in the network and the network has no direct control over how the applications perform their end-to-end adaptation. In cellular networks, application traffic is classified and marked at the packet gateway for the downstream traffic and at the UE for the upstream traffic to obtain the desired quality of service (QoS). The traffic classification can be based on policies derived via offline methods or can be based on online methods, e.g. utilizing deep packet inspection (DPI). Network elements can utilize certain packet fields (e.g. ECN bits) or selectively drop packets (e.g. employ active queue management) to signal traffic conditions (e.g. congestion) in the network as packets traverse the network. These explicit or implicit signals eventually propagate back to the traffic source through acknowledgments and/or measurement reports from the traffic sink, upon which the traffic source adjusts the sending rate at the session transport (e.g. TCP) or application layer.

The aforementioned mode of operation is seemingly straightforward but in practice it involves multiple control loops at different timescales operating under different objectives and constraints that often negatively interfere with each other. For instance, the end-to-end methods for available rate estimation can easily overshoot the instantaneously available rate in the network leading to large queuing delays. End-to-end methods can also miss instantaneous high rate transmission opportunities if they underestimate the available rate. Furthermore, providing a number of QoS classes and mapping individual traffic flows in a static manner onto these QoS classes either lead to over-provisioning or under-provisioning.

When there is a single class of URLLC applications that have predictable patterns (e.g. payload size and packet inter-arrival times) with well-characterized operational environment and communication constraints, one can engineer a network accordingly. Although some use cases satisfy such conditions, many URLLC applications exhibit more complex application traffic patterns and operational characteristics (e.g. see the listed requirements in Section 6.2). Further, in many cases the requirements are not absolute but conservatively estimated figures.

When NFV and SDN are brought into the picture, the overall latency, throughput and reliability are no longer simply a matter of how packets are routed or how link capacities are shared but also depend on how functional composition and configuration is done, where functional components are placed, and how computational resources are scheduled and consumed. Thus, interfacing between applications and network should have the agility to take advantage of such flexibility. Below, we summarize some considerations on how such interfacing can be done.

Figure 6.10 provides a high level view of interaction of application and access cloud that includes RAN, CN, and multi-access edge computing (MEC). Today, it is very common to have an application with client side code running on the user device and server side code running inside over-the-top (OTT) cloud. It is also very common for web services to offload their content to content delivery networks (CDNs) to improve latency, throughput, and scalability. However, CDNs themselves have no control over the access networks and cannot provide guaranteed services that URLLC applications would require. Thus, it is necessary that

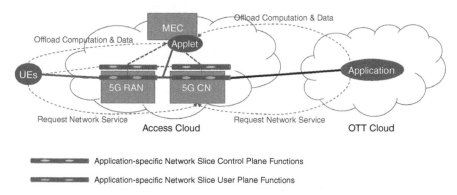

Figure 6.10 High level view on how UEs and applications utilize access cloud.

access clouds not only offer hosting services for application code and data but also network slices extending from the UE to the application code in order to guarantee end-to-end QoS.

In Figure 6.10, computation and data offloading can happen both from OTT cloud to MEC as well as from UEs to MEC [23]. The application logic that runs on the MEC side is labeled as *applet*, which could be deployed within lightweight containers. Note that in various scenarios the complete application stack might reside at the edge (i.e. in MEC and UEs). The application specific network slices are created both in CN and RAN with application-specific user plane as well as control plane functions. Common examples of control plane functions would be related to monitoring, pricing, mobility management, and slice configuration. Examples of UPFs include tunneling, encryption, traffic filtering, metering, marking, active queue management, caching, deduplication, transcoding, network coding, routing (multi-path, anycast, multicast), load balancing, etc.

Network slicing serves as the main abstraction that interfaces applications with the network. Unlike the case where one network carries all workloads, network slicing provides an explicit intent on communication services offered by the network and communication services requested by the applications. Thus, the objectives of the applications and the network can be aligned. In the current 5G spec (i.e. Release 15), only three broad slice types are defined, namely, eMBB, URLLC, and massive IoT. These are envisioned as quasi-static network services that are available in large service areas utilizing NFV in the core and slice-aware provisioning in the RAN. Thus, Service Level Agreements (SLAs) and capacities for each network slice are provisioned in a proactive manner with the mindset of traditional top-down network planning. Application-specific slicing that is imagined as a real-time interface between applications and the network is beyond the current 3GPP specification.

Figure 6.11 shows the benefit of providing application specific slicing. With pre-built generic slice types, a new type of URLLC application needs to be admitted into either a slice type that can provide performance that provides higher than targeted performance at a higher cost or lower than the targeted performance albeit at a lower cost. When the performance utility curve as a function of the amount of consumed resources is concave (e.g. with diminishing returns as in Figure 6.11), providing an application-specific slice can deliver just the right performance consuming the minimum amount of resources.

Network slices are created and modified based on the network service requests coming from UE or the server side of the application layer. The application programming interfaces (APIs) for network service requests might be intent-based APIs. Intent-based APIs do not specify how the network service is delivered and applications themselves are unaware of the network slices. One level of intent can be directly based on SLA targets and specify a request such as "provide a downlink

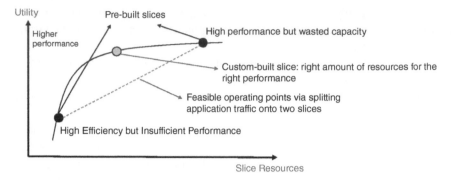

Figure 6.11 The trade-off between resource efficiency and performance in network slicing.

bandwidth of 1Gbps from Applet X to UE Y with a maximum packet latency of 5 ms". However, application level intents need not be intrinsically unique values, i.e. intents might change as context such as network conditions, pricing, location, customer preferences, etc. change. For more dynamic and iteratively refined service provisioning, the network should support APIs to provide real-time availability as well as pricing information of various SLA levels. Machine learning based algorithms can be used to steer the slice functionality and resources toward the desired level of service for the end users given a particular context.

For ultra-low latency, once the geographical proximity is shortened, the most critical aspect is the queueing delay. In cellular networks, it is typical for base stations to isolate traffic coming from distinct UEs and provide a fair scheduling across them. Thus, most queueing occurs at the bottleneck wireless link due to self-congestion, i.e. source rate exceeding what is actually available based on the presence of other flows and channel states. The network can provide resource allocation guarantees to network slices but cannot guarantee the actual transmission capacity realized over these resources as capacity is dictated by the channel states. However, the network has all the information available to forecast the available capacity per application flow and how it fluctuates. Feeding such information to applications via open (potentially slice-specific) APIs can eliminate almost all queueing delays while closely following the available capacity. As the slices are application specific, different applications are isolated and protected against each other at the slice level. Thus, misbehaving applications simply hurt themselves. For different network flows of the same application, the corresponding network slice instance should have the right end-to-end buffer isolation and fair scheduling among network flows within its realm in line with how rate estimations are performed.

One major concern against the paradigm shift of application-specific slicing is about the system complexity and its scalability as the number of potential applications can be tremendously high. Since mobile networks are fluid in nature with demand and wireless link capacities shifting in time and space, network slices also need to be fluid in terms of functional composition, placement, and resource allocation [24]. The introduction of micro-services and function as service concepts [25] into the CN and RAN infrastructure will enable fine-grain and fast resource allocation/deallocation of computing and memory resources. Complementing the fine-grain dynamic allocation of traditional network resources (e.g. link bandwidth, spectrum, power, beams, etc.), they have the potential to pave the way for application-specific slicing. As a stepping stone toward this paradigm shift, the telephone companies can initially focus on two pillars on top of their programmable cloudified networks: (i) delivering dynamically scalable network slices with fewer broad categories as defined in Release 15 and address spatio-temporal dynamics of demand as well as radio network capacity, and (ii) network automation based on data analytics and machine learning. Once these building blocks are in place, the network operators can naturally evolve their infrastructure to support application-specific slicing as monetization opportunities emerge with high-end applications.

6.6 Summary

URLLC is the new frontier for the wireless networks. New markets within the realm of public networks as well as in the context of private isolated networks is emerging. The success of URLLC depends on how network operators (in both public and private settings) can trade off capacity, delay, mobility, energy efficiency, and reliability in a dynamic manner based on the specific applications and network environment. Network slicing adopted in 5G networks makes it possible to pick different trade-off points for different use cases. As more and more services are being adopted, there is a need to provide a continuum of these trade-off points to most effectively utilize the system resources. We have presented how new architectural concepts and network programmability are taking the next generation networking many steps closer to that target.

References

1 FD.io (2017). The Vector Packet Processor (VPP) https://fd.io/vppproject/vpptech/ (accessed 27 November 2019)

2 3GPP TR 22.804 (2018). Study on Communication for Automation in Vertical Domains.

3 3GPP TR 22.827 (2018). 3rd Generation Partnership Project; Technical Specification Group Services and System Aspects; Study on Audio-Visual Service Production (Release 17).

4 3GPP TS 23.251 (2018). 3rd Generation Partnership Project; Technical Specification Group Services and System Aspects; Network Sharing; Architecture and Functional Description (Release 15).

5 3GPP TS 23.501 (2018). 3rd Generation Partnership Project; Technical Specification Group Services and System Aspects; System Architecture for the 5G System (Release 15).

6 3GPP TS 23.502 (2018). 3rd Generation Partnership Project; Technical Specification Group Services and System Aspects; Procedures for the 5G System; Stage 2 (Release 15).

7 3GPP TS 28.531 (2018). 3rd Generation Partnership Project; Technical Specification Group Services and System Aspects; Management and Orchestration; Provisioning (Release 15).

8 Song, H. (2013). Protocol-oblivious forwarding: unleash the power of SDN through a future-proof forwarding plane. *Proceedings of the Second ACM SIGCOMM Workshop on Hot Topics in Software Defined Networking (HotSDN '13)*. New York, NY, USA: ACM, pp. 127–132.

9 Bosshart, P., Daly, D., Gibb, G. et al. (2014). P4: programming protocol-independent packet processors, ACM SIGCOMM. *Computer Communication Review* 44 (3).

10 Bianchi, G., Bonola, M., Capone, A., and Cascone, C. (2014). OpenState: programming platform-independent stateful openflow applications inside the switch. *Computer Communication Review* 44 (2): 44–51.

11 Lin, Y., Kozat, U.C., Kaippallimalil, J. et al. (2018). Pausing and resuming network flows using programmable buffers. *Proceedings of the Symposium on SDN Research (SOSR '18)*. Los Angeles, CA, USA: ACM.

12 ETSI (2014). Network Functions Virtualisation (NFV); Management and Orchestration. ETSI GS NFV-MAN 001 V1.1.1 (2014-12).

13 Intel (2011). PCI-SIG SR-IOV Primer: An Introduction to SR-IOV Technology. https://www.intel.com/content/dam/doc/application-note/pci-sig-sr-iov-primer-sr-iov-technology-paper.pdf (accessed 18 October 2019).

14 Bernal, M.V., Cerrato, I., Risso, F., and Verbeiren, D. (2016). A Transparent Highway for inter-Virtual Network Function Communication with Open vSwitch. *Proceedings of the 2016 ACM SIGCOMM Conference (SIGCOMM '16)*. New York, NY, USA: ACM, pp. 603–604.

15 Intel (2013). DPDK: Data Plane Development Kit. http://dpdk.org.

16 Panda, A., Han, S., Jang, K. et al. (2016). NetBricks: taking the V out of NFV. *Proceedings of the 12th USENIX Conference on Operating Systems Design and Implementation (OSDI'16)*. Berkeley, CA, USA: USENIX Association, pp. 203–216.

17 Han, S., Jang, K., Panda, A. et al. (2015). SoftNIC: A Software NIC to Augment Hardware. Technical Report UCB/EECS-2015-155, EECS Department, University of California, Berkeley, USA.

18 Rizzo, L. (2012). Netmap: a novel framework for fast packet I/O. *Proceedings of the 2012 USENIX Conference on Annual Technical Conference (USENIX ATC '12)*. Berkeley, CA, USA: USENIX Association, p. 9-9.

19 AWS (2018). Amazon EC2 F1 Instances. https://aws.amazon.com/ec2/instance-types/f1 (accessed 18 July 2018).

20 Yi, X., Duan, J., and Wu, C. (2017). GPUNFV: a GPU-Accelerated NFV System. *Proceedings of the First Asia-Pacific Workshop on Networking (APNet '17)*. New York, NY, USA: ACM, pp. 85–91.

21 Kubernetes (2018). Learn Kubernetes Basics. https://kubernetes.io/docs/tutorials/kubernetes-basics (accessed 24 July 2018).

22 Auge, J., Carofiglio, G., Grassi, G. et al. (2015). Anchor-less producer mobility in ICN. *Proceedings of the 2nd ACM Conference on Information-Centric Networking (ICN'15)*. New York, NY, USA: ACM, pp. 189–190.

23 Satyanarayanan, M., Bahl, P., Caceres, R., and Davies, N. (2009). The Case for VM-Based Cloudlets in Mobile Computing. *IEEE Pervasive Computing* 8 (4): 14–23.

24 Leconte, M., Paschos, G.S., Mertikopoulos, P., and Kozat, U.C. (2018). A resource allocation framework for network slicing. *IEEE Infocom 2018*, Hawaii, USA (April 2018).

25 Hendrickson, S., Sturdevant, S., Harter, T. et al. (2016). Serverless computation with OpenLambda *8th USENIX Workshop on Hot Topics in Cloud Computing (HotCloud'16)*, Denver, CO, USA (June 2016).

7

Edge Cloud

An Essential Component of 5G Networks

Christian Maciocco[1] and M. Oğuz Sunay[2]

[1] *Intel Corporation, Hillsboro, OR, USA*
[2] *Open Networking Foundation, Menlo Park, CA, USA*

Abstract

This chapter first describes the fundamentals of edge cloud with respect to 5G networking and the associated on-going mobile network transformation. It then describes Software defined networking, its evolution from OpenFlow to P4, and network function virtualization and its implications for realizing disaggregated, control plane – user plane separated mobile core. The chapter introduces an open source platform for mobile core, namely Open Mobile Evolved Core, and discusses its evolution path. Next, it describes the disaggregation and software defined control of radio access networks, and describes the rise of white-box hardware for computation, storage, and networking, and how modern networks will take advantage of cloud technologies at the edge to meet the end devices' ever increasing demands. The chapter further provides a description of end-to-end, software defined network slicing, including programmatic slicing of the RAN and provides insights into how network slicing is related to the edge cloud.

Keywords *5G networking; edge cloud deployment options; mobile network transformation; network function virtualization; Open Mobile Evolved Core; servicebased architecture; software defined networking; softwaredefined disaggregated radio access networks whitebox solutions*

5G Verticals: Customizing Applications, Technologies and Deployment Techniques, First Edition.
Edited by Rath Vannithamby and Anthony C.K. Soong.
© 2020 John Wiley & Sons Ltd. Published 2020 by John Wiley & Sons Ltd.

7.1 Introduction

This chapter is divided into six parts and each part is an attempt to provide readers with an insight into the past, present and the future of networking.

Part I describes the fundamentals of edge cloud with respect to 5G networking and the associated on-going mobile network transformation.

Part II describes Software defined networking (SDN), its evolution from OpenFlow to P4 [1], and network function virtualization (NFV) [2] and its implications for realizing disaggregated, control plane–user plane separated mobile core. An open source platform for mobile core, namely Open Mobile Evolved Core (OMEC), is introduced and its evolution path is discussed. The section also describes the features of high-volume-servers to perform associated packet processing either in bare-metal or in an orchestrated cloud environment. Challenges on containerization and networking are discussed.

Part III describes the disaggregation and software defined control of radio access networks (RANs). The disaggregation of the base station into radio unit (RU), distributed unit (DU) and central unit (CU) is overviewed. This part also introduces how software defined controllability of base station nodes is possible and introduces the associated architecture.

Part IV describes the rise of white-box hardware for compute, storage, and networking, and how modern networks will take advantage of cloud technologies at the edge to meet the end devices' ever increasing demands. This part also provides examples of Open Platforms such as Open Networking Foundation's (ONF's) [3] Central Office Re-Architected as a Datacenter (CORD) [4], Open Networking Automation Platform (ONAP) [5], open radio access network (O-RAN) [6, 7] and Facebook's Telecom Infrastructure Project (TIP) [8] and Open 5G NR [9].

Part V attempts to answer the question, "where is the edge?" and provides insights into how operator, over-the-top (OTT), and end-user workloads can be placed in the distributed network cloud paradigm. This part also provides an overview of a demonstration as a concrete example.

Part VI provides a description of end-to-end, software defined network slicing, including programmatic slicing of the RAN and provides insights into how network slicing is related to the edge cloud.

Finally, we conclude the chapter with a brief summary.

7.2 Part I: 5G and the Edge Cloud

Cellular communication has been shaping society for the last 40 years. Approximately every 10 years we have a new generation of cellular standards that take us to the next level. Until recently, this evolution has focused on first enabling and then scaling two distinct services: voice and broadband data. The resulting

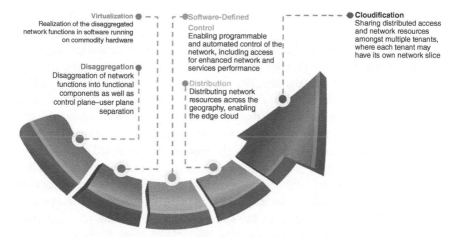

Virtualization ● — ⌐ ┌ ●Software-Defined ┌ ─ ─ ─ ─ ─ ─ ●Cloudification
Realization of the disaggregated ┊ ┊ Control ┊ Sharing distributed access
network functions in software running ┊ ┊ Enabling programmable ┊ and network resources
on commodity hardware ┊ ┊ and automated control of the┊ amongst multiple tenants,
┊ ┊ network, including access ┊ where each tenant may
┊ ┊ for enhanced network and ┊ have its own network slice
Disaggregation ● ┊ ┊ services performance ┊
Disaggreation of network ┊ ┊ ┊
functions into functional ┊ ┊ ●Distribution ┊
components as well as ┊ ┊ Distributing network ┊
control plane–user plane ┊ ┊ resources across the ●
separation ┊ ┊ geography, enabling ┊
┊ ┊ the edge cloud

Figure 7.1   Mobile network transformation.

massive explosion of data usage has resulted in a bleak economic reality for tele-communication operators. Their average revenue per user (ARPU) have been steadily declining over the last number of years. This is mainly because the industry is approaching the limits of both consumer value creation and current network capabilities. 5G is being architected to counter this unsustainable progression. For the first time since the inception of cellular communications, voice and broadband data are not the only use cases for which we optimize the cellular network. Two new, additional broad categories for new revenue generation are ultra-reliable, low latency communications and support for massive Internet of Things (IoT).

To enable these new use cases a transformation of the network architecture is necessary. Building on the pillars of virtualization, disaggregation, software defined control, and cloud principles over a geographically distributed compute, storage and networking paradigm, this transformation aims to bring the operators agility and economics that are currently enjoyed by cloud providers. The aim is to allow rapid creation of new connectivity services while lowering the necessary capital and operational expenditures. The components of this transformation are summarized below and illustrated in Figure 7.1.

Virtualization: The first step in the network architecture is NFV. With NFV, the traditionally vertically integrated, proprietary realizations of network functions are realized in software to run on commodity hardware.

Disaggregation: Next in the transformation is the disaggregation of the network functions. This disaggregation takes place in two dimensions: functional disaggregation and control plane–user plane disaggregation. While the former paves the way toward microservices-based, cloud-native realization of the network functions, the latter enables programmatic, software defined control of access, connectivity, and networking and independent scaling of the user and control planes, an essential feature for efficient resource utilization when different use cases with significantly different requirements share a common infrastructure.

Software defined control: The network architecture transformation requires software defined control so that new access, connectivity and networking services can be rapidly created, and dynamically lifecycle managed. Furthermore, SDN-based 5G network allows for more efficient network resources management and control using advanced applications that are potentially machine learning driven.

Distribution: A 5G network where infrastructure and resources are shared among different use cases and business verticals require a distributed architecture. Specifically, in addition to access resources, compute, storage and networking resources need to be distributed across the geography so that services that require low latency, high bandwidth and traffic localization can be offered where needed.

Cloudification: The last step in the transformation is cloudification. This is enabled by network slicing and multi-tenancy. With cloudification, the distributed network resources are allocated to different use cases as well as tenants in an elastic and scalable manner. Further, the utilization of these resources by the use cases and tenants are tracked for monetization.

Let us discuss how this network architecture transformation is taking place. The current mobile network architecture can be illustrated in a simplified form as in Figure 7.2. In this architecture, geographically distributed base stations provide coverage and mobile connectivity for subscribers' user equipment (UE). Traffic from the base stations are aggregated toward centralized operator locations, where vertically integrated, proprietary realizations of mobile core network functions as well as operators' own applications such as VoLTE, texting, and streaming reside. A subset of these locations also acts as the ingress/egress locations for the cellular network toward the Internet toward the public clouds, where the majority of the popular OTT applications reside.

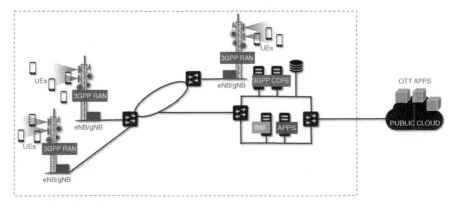

Figure 7.2 Current mobile network architecture.

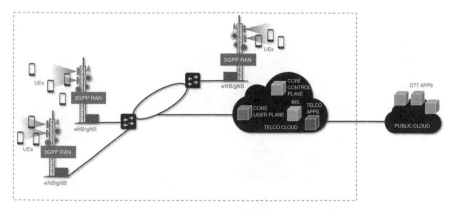

Figure 7.3 First step in transformation of the mobile network architecture: disaggregation and virtualization of network functions.

Toward 5G, the transformation of the network architecture has already started for many of the Tier-1 operators with virtualization and disaggregation of the network functions that reside in the operators' centralized locations, as illustrated in Figure 7.3. The figure shows the user and control planes of the core as single components for concise illustration – in reality, functional disaggregation is also present for both today and they are composed of multiple components. Virtualization and disaggregation have prompted the transformation of these centralized operator locations effectively into cloud platforms so that the virtualized network function workloads can share the pool of compute, storage and networking resources.

This first step, while beneficial in lowering the operator expenditures when deploying 5G networks, is not sufficient in realizing the different connectivity options necessary to enable new revenue generating business verticals. This requires a distributed network paradigm where edge plays a crucial role.

There are three broad reasons why edge cloud is important. Werner Vogels, the CTO of Amazon, describes these reasons as "laws" because they will continue to hold even as technology improves [10].

The first law is the *Law of Physics*. The new functions that are to be built toward satisfying the new business verticals usually need to make the most interactive and critical decisions quickly. One example of such a business vertical is Industrial Automation where safety-critical control requires low latency. If such control were to be realized in the cloud, we know from basic laws that it takes time to send the data generated at the edge to the cloud, and a lot of these services simply do not have the tolerance for this delay. The second law is the *Law of Economics*. In 2018, for the first time in history, we observed more IoT devices connected to the network than people. These devices are starting to generate a significant amount

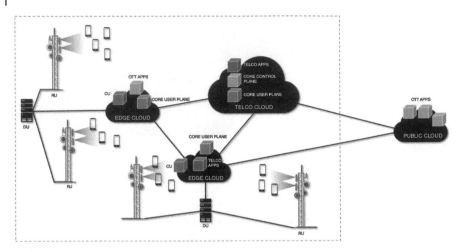

Figure 7.4 Next step in transformation of the mobile network architecture: mobile edge cloud.

of data at the edge. The goal is to unlock the value from this data, which is done by processing it. However, the increase in the data generated at the edge is faster than the increase in available backhaul bandwidth. Then, local aggregation, filtering and processing of data at a location close to where it is generated will allow the backhaul networking capacity demand not to grow out of bounds. The third law is the *Law of the Land*. In some business verticals, such as healthcare, it is required to keep data local and isolated by law. Further, increasingly, more governments impose national data sovereignty restrictions to keep their data localized. Also, some of the business verticals, such as offshore oil rigs, require localized services at remote locations where connectivity to the outside world is not necessary, or is not even possible.

A distributed 5G network architecture with edge clouds is illustrated in Figure 7.4. The first benefit the edge cloud brings is that it allows for the virtualization and disaggregation of the base station. Following an O-RAN architecture [11], we foresee three components: RUs at antenna sites, DUs at cell sites, and CUs that are virtualized and pooling DUs and RUs and are hosted at the edge clouds. Further details of RAN virtualization and disaggregation can be found in Part III. In addition to hosting the CU and thereby enabling RAN virtualization and disaggregation, the edge cloud is also an enabler of low latency, high bandwidth, and localized services. Then, the edge needs to be a place where local traffic break-outs are enabled. This means, as illustrated in Figure 7.4, the virtualized user plane component(s) of the mobile core network needs to be instantiated at the edge. Toward this end, we also need to host the operator (telco) applications that require low latencies and or traffic localization at the edge. Then, depending on which applications are instantiated at specific edge cloud locations, operators

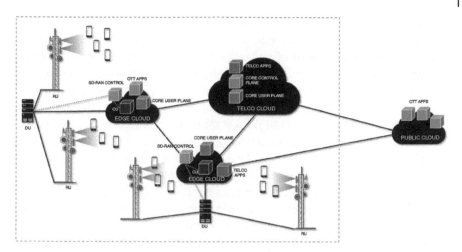

Figure 7.5 Following step in transformation of the mobile network architecture: software defined RAN Control.

will be able to offer specialized services new business verticals. Via network slicing, multiple such verticals could potentially share the same edge cloud. If the operators are open to having a multi-tenant edge cloud, OTT applications or components of such applications can also be hosted at the edge, benefitting from close proximity to end-users and the presence of local-break outs.

We listed software defined control as a fundamental component of the network transformation. An important aspect of this control is the control of the access – the last mile. When such control is employed, the associated controller will be making decisions and controlling the CU and DU components in response to time-varying RAN characteristics in near real-time. This necessitates that the associated RAN controller be also instantiated at the edge as illustrated in Figure 7.5.

Cloudification is the elastic and scalable management of a common pool of distributed network resources (access, compute, storage, and networking) so that a common infrastructure may be efficiently and profitably shared among multiple business verticals and tenants and is the last component of the 5G network transformation. Network slicing is an essential enabler for cloudificaton. We briefly describe network slicing in Part VI.

7.3 Part II: Software Defined Networking and Network Function Virtualization

In less than a decade, cloud computing-based services and applications, enabled by high-volume-server's ability to process networking workloads, increased broadband connectivity and mobile devices or smartphone capabilities, have resulted in many new big data opportunities in different industries. Our daily activities are largely engrained with mail, storage and collaboration tools, music services, photo or video sharing, and social networking, which are available to us wherever we go. Operators see this opportunity to turn cloud-based approaches to their advantage and implement new architectures that provide network efficiency, quality of experience (QoE), support scalable applications for large number of end-users and shorten the time-to-market for innovative services.

All these are possible thanks to network programmability and a common delivery platform which was introduced by a new architectural approach that leverages the availability of massive compute resources, called SDN. Highly complementary to SDN is NFV which does not necessarily depend on SDN. NFV is an alternative design approach for building complex applications, specifically in the telecommunications industries, by virtualizing different new service functions into building blocks which can be stacked or chained together.

7.3.1 Rise of SDN

Several years ago, a number of university research groups at UC Berkeley and Stanford University started with the idea of separating the control plane from the data plane in computer networking architecture [12]. They experimented with this by disabling the distributed control plane in Ethernet switches and managed the flow of streams through these switches via centralized software control. The SDN approach was claimed to have a quick and reliable deployment of production network, alleviating time-consuming setup and providing error prone device-by-device connection. The OpenFlow specification and later the P4 programming language [13] from ONF were some of several "SDN protocols" that sprang out of this effort.

The control plane in SDN is moved into a high-powered central computing environment to achieve a centralized view and control of the network. An SDN controller is a middleware software entity which is responsible for the new centralized control plane. It runs the algorithms and procedures to calculate the paths for each flow of packets and communicates this forwarding information to other networking devices. It also supports applications running on top of application programming interfaces (APIs) simultaneously so that the applications can discover and request network resources, either controlled by operations staff or fully automated.

7.3.2 SDN in Data Centers and Networks

In recent years, the rise of virtual machines (VMs) has led to improved server efficiency; however, this has resulted in increased complexity and a huge number of network connections due to the large number of VMs per server. The next step for SDN was to be utilized in data centers. Centralized SDN controllers have created value by analyzing the entire data center network as an abstracted and virtualized representation. These controllers are capable of simulating any network changes ahead of time, and therefore, automatically configuring and preparing different switches when the scenario was activated.

More recently, architects have also been experimenting with SDN in the wide area network. The most well-known experiment to date is the deployment of central SDN control on Google's G-scale network [14], as well as custom-built open switches connecting its data centers around the world. The following benefits were claimed:

- Faster with higher deterministic convergence when compared with the distributed control plane.
- Increased operational efficiency, with links running at 95 versus 25%.
- Fast innovation support through the network simulation in software before production service deployment.

With the advent of SDN and the separation of control and data plane came the opportunity to define, and program, the data plane processing stages of packet forwarding devices, whether implemented in a switch, NIC, FPGA, or CPU. P4 [13] is a domain specific high-level programming language supporting constructs optimized around packet processing and forwarding for protocol independent switch architecture (PISA). A combination of PISA-based devices and the P4 language will allow the programmer to achieve device and protocol independence and enable field reconfiguration:

- *Device independence*: P4 programs are designed to be implementation independent and can be compiled on any targets, e.g. CPUs, FPGAs, switches, etc. The programmer defines the processing that will be performed on the packets as they progress through the various stages of the packet forwarding pipeline.
- *Protocol independence*: With P4 there is no native support for any protocols as the programmer describes how the incoming packets will be processed through the stages of the PISA device. A device be it a switch, a CPU, or an FPGA is not tied to a specific version of the protocol as the processing for any protocol will be defined programmatically using the P4 language.
- *Field reconfiguration*: Due to the target device and protocol independence and the abstract language, P4-based PISA devices can update their behavior as the need to process different fields, different protocol version, or take different actions evolves over time. For example, when a protocol version changes with new and updated field a simple P4 software update taking into account the new fields is developed, recompiled for the specific target device and downloaded to the device enabling field reconfiguration. There is no need to wait for a new spin of the silicon to support a protocol update.

The P4 language is evolving and currently P4_16 [15] is defined as a stable language definition with long-term support. P4_16 supports ~40 keywords and has the following core abstractions: "Header Types" to define the header's fields, "Parser" to instruct how to parse packet headers, "Tables" containing state associating user-defined keys with actions for the "Match-Action" command where "Match" create a lookup key, perform lookup using, e.g. exact match, longest prefix match or ternary match as defined by the user, and then "Action" describing what to do on the packet when a match occurs. "Control Flow" specifies in which order the match-action will be performed. Then "External Objects" define registers, counters, etc. and "Metadata" defines data structure/information associated with each packet through the processing stages of the PISA-based device. Figure 7.6 represents a high-level view of a P4 pipeline on a PISA-based device.

We can view P4 along with PISA-based devices as enabling the programmer to do a top-down approach where the programmer will instruct the device what to do with a packet whereas before P4 and PISA-based devices it was more of a bottom-up approach where the devices dictated how the packets would be processed. There are some limitations as the programmer needs to know how many tables in order to perform match/action processing on the packets but P4 and PISA-based devices enable significantly higher programmability than previously. P4 and PISA-based devices enable new usage models, e.g. in-band telemetry on a per packet basis, or performed network functions in the fabric itself.

In Figure 7.7 the communication between the control plane and data plane can be handled using P4 Runtime [15, 16] that can be viewed as an open switch API for control plane software to control the forwarding plane of a packet processing device pipeline, the device either being a programmable switch ASIC, an FPGA, a

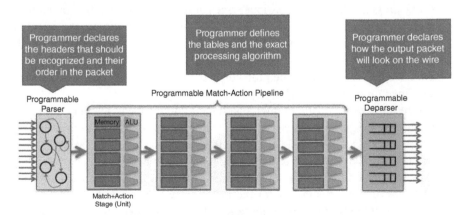

Figure 7.6 PISA architecture and associated P4 pipeline [13].

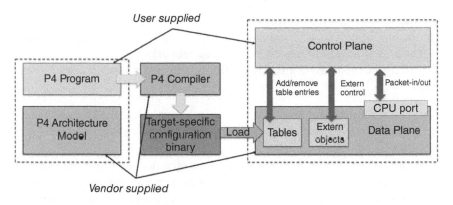

Figure 7.7   P4 workflow [13].

NIC or a software switch running on an x86 platform. P4 Runtime is agnostic to where the control plane resides, whether on a local switch operating system or on a remote control plane on x86 servers, and solves issues that could not be addressed by previous attempts at defining an open switch API, mainly OpenFlow. Over the last 10 years the industry has learned from experience that OpenFlow suffered from being complex, hard to extend, and ambiguous in its behavior as it only mandates the fields in a packet that should be matched upon but did not always define the actions to be performed on a match leading to different interpretations by different vendors.

As an example, in Figure 7.7 a P4 program would define the processing pipeline stages of the data plane and the P4 compiler would generate the scheme to be used by the P4 Runtime APIs to add/remove entries in/from the tables at runtime. If the data plane is programmable the developer may extend the P4 program to add new tables to be controlled a runtime, for example to support a new protocol, and the compiler will extend the schema so that the new tables can be controlled through the P4 Runtime API.

7.3.3   Network Function Virtualization

Including the WAN as a component of cloud-based services and combining the network-enabled cloud and service providers' SDN capabilities enables the network operators to empower network virtualization and unlock their full potential, including:

- Creating a layered/modular approach to allow innovation at different and independent layers/modules.
- Separating the data plane from the control plane.

- Providing a programmable network through the deployment of a network abstraction layer for easier operation/business application development without understanding the complexity of various networks.
- Providing the capability of creating a real-time and centralized network weather map.

Operators can now simplify their networks, by removing the complexities of their network topology and service creation, and accelerating the new service process creation and delivery by incorporating NFV.

NFV aims to transform the way that network operators architect networks by implementing the network functions in software that can run on any industry standard high volume server which can be instantiated in any locations in the network as needed, without requiring the installation of new equipment. With NFV the operators' or service providers' deployed equipment can take multiple personalities, e.g. being a router one day, a firewall or other appliance the next.

By network virtualization, the network can be mapped and exposed to logical abstractions instead of a direct representation of the physical network. Therefore, logical topologies can be created to provide a way to abstract hardware and software components from the underlying network elements and to detach control plane from forwarding plane capabilities and support the centralization of control, e.g. SDN. In addition, the network virtualization can be consolidated in computing and storage virtualization to further provide:

- Replicating cloud IT benefits inside their own telecom networks.
- Service velocity and on-demand provisioning.
- Quicker time to new feature and services driving new revenues.
- Third party apps on deployed infrastructure.
- Operational efficiency.
- Abstract complexity in the network, improved automation and agility.
- Simpler multi-vendor network provisioning.
- Lower operating expense with unified view (management/control) of multi-vendor, multilayer.

As the mobile wireless core evolves toward 5G with one of its goals to support edge networking and have a solution available on off-the-shelf high volume platforms based on open-source software a set of partners including operators, software systems providers and merchant silicon providers used operators real 4G/LTE traffic for a detailed performance analysis of the system's components of 4G/LTE wireless mobile core [17] to evaluate the benefits of the architectural changes brought by SDN and disaggregation [18], NFV, as well as Cloud Native operation.

A key goal of this analysis is to understand the impacts of the interaction of control and data plane messages on the Service Gateway (SGW) and Packet

Gateway (PGW) of the mobile core, shown in Figure 7.8, to see what benefits disaggregation would bring and how to partition the system to operate in a Cloud Native environment.

A detailed analysis of the interactions between the MME and SGW showed that ~41% of the control plane messages on the S1-MME interface are forwarded to the SGW's S11 interface [18]. By removing single transaction messages such as "Idle-to-Connect" or "Connect-to-Idle" the remaining multi-transaction messages account for 33% of transferred messages from the MME to the SGW. Approximately 18% of these messages are then forwarded from the SGW to the PGW on the S5/S8 interface. These results show that the SGW, which also handles the S1-U data plane interface is the system's component where control messages processing will have the biggest impact on the data plane processing, especially as the load on the data plane increases. Figure 7.9 shows graphically that 41% of control plane messages impact the SGW data plane.

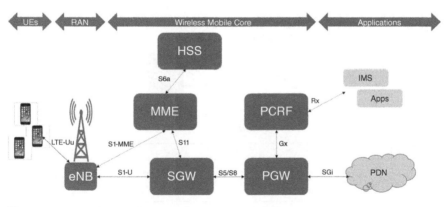

Figure 7.8 Simplified 4G/LTE architecture.

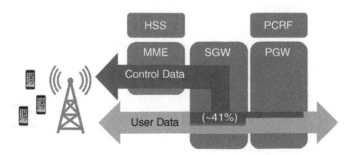

Figure 7.9 Pictorial of control plane message impacts on the data plane in traditional architecture.

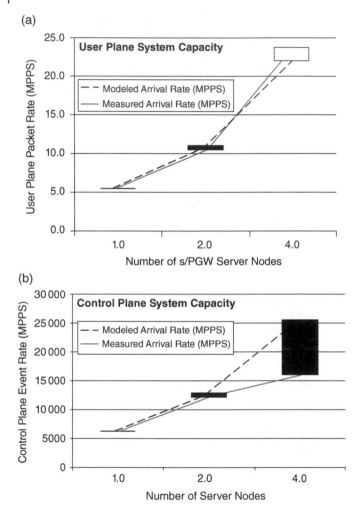

(a)

(b)

Figure 7.10 (a) User plane and (b) control plane queue flooding [17].

The graphs in Figure 7.10 show the linear throughput scalability of the data plane and the non-linear event rate scalability of the control plane when operating the system in a cluster. In Figure 7.10a the S/P-GW data plane scales linearly with throughput going from ~5 MPPS with one server node, the thin rectangle on the left-hand side of the graph, to ~10 MPPS with two servers, the dark rectangle in the middle of the graph, to ~20 MPPS with four server nodes, the white rectangle at the top of the graph. Meanwhile the control plane events processing does not scale linearly as shown in Figure 7.10b. The control plane scales from ~6.5 MPPS on one server, to ~13 MPPS on two servers, the dark rectangle in the middle of the graph. Then, while the simulation shows that we should reach ~26 MPPS using four servers, reaching the top of the black rectangle, we only reach ~16 MPPS using real measurements, the bottom of the black rectangle.

This limitation of control message processing on the current architecture will only become worse moving forward with the expected growth of mobile devices and usages on 5G networks. Some of these usage models will have heavy signaling traffic with limited and sparse data traffic, e.g. IOT Meters while others might be heavy on data traffic, c.g. IOT Security Cameras. To support the growth of devices and usage models, the wireless mobile core needs to scale efficiently, but today's SGW and PGW scale both the control plane and data plane simultaneously. If a usage model only requires scaling, we will have an over-provisioned data plane or vice versa. This is not the optimal way to adapt system components to traffic requirements as we want to have the component under load scales for what needs to be scaled. For the SGW or PGW we want the control plane to scale independently from the data plane based on usage requirements.

The networking industry is undergoing architectural transformation with SDN principles enabling the disaggregation of the control and data plane [1] as well as efforts in 3GPP defining the Control and User Plane Separation (CUPS) [17] for 5G networks (although CUPS defined for Release 14 could be implemented with 4G/LTE networks). NFV [2] is another transformation enabled by the significant progress made by standard high volume servers to process networking workloads.

Following these architecture principals the virtual EPC SGW and PGW were re-architected and the resulting implementation is not simply a port of a hardware functionality to a software one running in a bare-metal process, a VM or a container but instead a carefully thought out architecture, design and implementation to create an efficient and high performance SGW and PGW. Figure 7.11 shows the

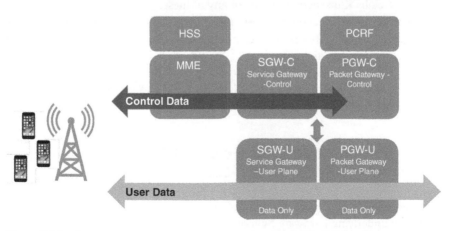

Figure 7.11 Control and data plane separation following 3GPP CUPS architecture.

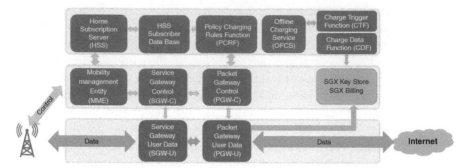

Figure 7.12 Open Mobile Evolved Core.

updated architecture where the SGW and PGW have a data plane following a match/action semantic with optimized data structures enabling fast and efficient lookup of a large number of keys in the millions. Figure 7.11 shows the disaggregated wireless mobile core with the SGW-C and PGW-C, the control plane, separated from the SGW-U and PGW-U data planes. The reference implementation supports various communication mechanisms between the control plane and data plane, e.g. using OpenFlow v1.3 with a significant number of vendor's specific extensions as OpenFlow has short comings to carry all the necessary information, e.g. as related to billing with this version. A proprietary transport mechanism is also used until the 3GPP CUPS Sxa (SGW-C to/from SGW-U), Sxb (PGW-C to/from PGW-U) and Sxc (TDF-C to/from TDF-U) interfaces are fully ratified.

A reference implementation developed by industry partners called OMEC [19] has been released by ONF and is illustrated in Figure 7.12. Each component of the reference architecture can run as a bare metal process, VM, Docker container orchestrated by Kubernetes or a mix of any of these.

In the OMEC reference implementation, the SGW-C/U and PGW-C/U can be configured to operate as a System Architecture Evolution (SAE) Gateway, i.e. with S-GW and P-GW functionality.

Open source reference implementations of wireless core in addition to OMEC exist in the mobile ecosystem. Two distinct examples are OpenAirInterface [20] and Facebook Magma [21].

7.4 Evolving Wireless Core, e.g. OMEC, Towards Cloud Native and 5G Service-Based Architecture

Key promises of NFV were the decoupling of the virtual network function (VNF) from the underlying hardware by moving the functionality from hardware to software environment, e.g. VM, containers, etc. and to enable such VNF to dynamically scale based on workload requirements. This scaling could be scaling-up or vertical scaling where additional resources like CPU cores, NIC queues, etc. are added/removed to/from the VNF to increase its performance, or scaling-out or horizontal scaling where instances of the VNF are dynamically instantiated. A combination of vertical and horizontal scaling is also possible where the VNF scales up to a predefined threshold then the system orchestrator based on collected VNF's key performance indicator metrics will launch a new instance of the VNF thus scaling horizontally, migrating the concept of VNF to Cloud native network function (CNF).

As OMEC components run either in bare-metal, VM or container environment, the next step was to take these components into an orchestrated environment toward being Cloud Native. What does it mean to be Cloud Native? Being Cloud Native has few tenets, among them microservices, containers, continuous integration/continuous delivery (CI/CD), and devops. Micro-services-based architecture, usually deployed as containers, is where applications are a collection of loosely connected services where each service can scale independently. Containers package code and its dependencies allowing the applications to run reliably from one environment to another. Containers share a single operating system instance but each one has its own file system and system resources. CI/CD enables continuous integration and development following devops principles of lean and agile development from usually an open source developer community environment, and tests enabling rapid changes and deployment for the VNF.

In the meantime as Cloud Native architecture is taking hold in the cloud and data center industry, 3GPP is defining the upcoming 5G architecture for the wireless core, the 5G service-based architecture (SBA) where components of this architecture are microservices based, i.e. containers based. Having the wireless core transition to a Cloud Native architecture will enable horizontal scaling of individual components of the 5G SBA control plane, scalability triggered by reaching thresholds set by operators for specific components, e.g. thresholds based on the number of active sessions or loads on specific components. To enable Cloud Native scaling the various components should be stateless, i.e. the particular states of the connection are saved/stored outside of the component itself in distributed database or key/value store, also providing a fault-tolerant and fault-recovery mechanism service. The 4G/LTE Release 14 architecture defining CUPS does not lend itself well to be Cloud Native without a major restructuring of the architecture and software which has been addressed with the 5G SBA.

The OMEC reference implementation will evolve toward a Cloud Native architecture over time as it transitions toward a 5G Core network as shown in Figure 7.13 with a subset of 3GPP SBA interfaces.

To deploy and operate a Cloud Native solution we need an orchestrator to manage the lifecycle and run-time operation of the various containers/microservices. Kubernetes (K8s) [22] is an open source orchestrator for automated deployment

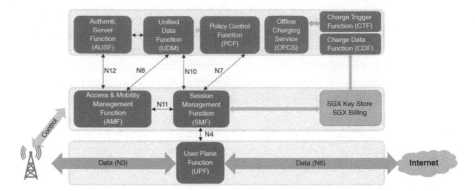

Figure 7.13   OMEC evolution toward 5G Core (subset of interfaces shown).

and lifecycle management of PODs/containers. A POD in a K8s environment has one or more containers, tightly coupled from an application point of view, sharing resources, and representing the unit of scheduling for the K8s scheduler. K8s recognizes and provides services, e.g. Domain Name System (DNS), to a single POD network interface created using an abstraction called Container Network Interface (CNI). Multiple CNI plugins exist in the industry and among them some enable support of multiple network interfaces, e.g. Multus [23], as the S-GW and P-GW components are using multiple network interfaces (S1U, SGi, Control Plane). A key issue with multiple network interfaces is to discover these interfaces network address information, as K8s only knows one POD network address. To enable service discovery for networks/interfaces not known to K8s and managed by K8s we use a key/value store service, e.g. Consul [24]. Over time K8s will evolve and maybe natively support multiple network interfaces per POD.

Figure 7.14 shows two K8s PODs, each with a single container for control and data plane functionality in a single node, but could be in separate nodes as well. The control plane just uses one interface eth0, known to K8s and for which K8s provides DNS and other services. The data plane uses three network interfaces, eth0 recognized by K8s and S1U and SGi ones whose IP addresses must be registered, discovered, and managed through a database/key value service, e.g. through Consul. The figure also shows the use of the SR-IOV CNI plugin in order to support SR-IOV to transfer the data directly from the NIC device into the container networking stack. An alternate solution is to have a single network interface where in the data plane case the traffic on the S1U and SGi interfaces would be demultiplexed in software thus exposing just one interface managed by K8s.

To achieve high performance while the VNFs operate in an orchestrated containers environment the system needs to be tuned for performance. Intel published a white paper on Enhanced Platform Awareness [26] in K8s. Critical system configuration is required to achieve high performance, among them (i) VNF

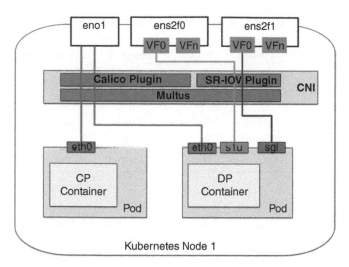

Figure 7.14 Kubernetes and S/P-GW C and U networking [25].

pinning to a specific CPU core, (ii) using huge memory pages, and (iii) single root input/output virtualization (SR-IOV) for I/O:

i) The VNFs need to be pinned to one or more CPU core(s) and these cores need to be isolated from the operating system so no other tasks than the VNFs processing are scheduled on these cores. Kubernetes offers a native CPU manager allowing the user to specify the resources, e.g. CPU core(s), memory, etc. dedicated to each POD. The user can specify the POD QoS class as *Guaranteed*, which guarantee an isolated deployment where the POD is the only scheduled resource on the specified CPU core.

ii) Huge pages are a large area of memory, e.g. 1 GB, compared with traditional memory page size of 4 kb. For optimal performance DPDK-based VNFs allocate memory for packet descriptors and buffer from huge pages. When using huge pages, fewer memory pages are needed thus reducing the number of TLB (Translation Lookaside Buffer) misses when a memory page is not found in the cache. K8s supports the allocation of huge pages by application in a POD.

iii) SR-IOV for VNFs requiring high-throughput, low latency I/O. A NIC supporting SR-IOV exposes independent virtual functions (VFs) for a physical function (PF), i.e. the physical interface and its control of the device. VFs can be configured independently with their own virtual MAC address and IP address. K8s SR-IOV support is implemented as a CNI plugin attaching the VF to the POD, thus transferring the Rx/Tx data directly from the POD to the device enabling the POD to achieve throughput performance near native bare metal process.

Table 7.1 Performance achieved by a user-space DPDK-based application [25].

Test	Usr Sp Drv	Pinning	Huge	Packets/s[a]	(with noise)
Native	Yes	Yes	Yes	1550K	*(1100K)*
Kubernetes	Yes	Yes	Yes	1450K	*(1150K)*
Kubernetes	No	Yes	Yes	750K	*(650K)*
Kubernetes	Yes	No	Yes	1450K	*400K*
Kubernetes	Yes	Yes	No	1200K	*(1100K)*

[a] 50K Granularity (1 CPU Core).

Table 7.1 shows the performance achieved by a user-space DPDK-based application (Usr Sp Drv) versus a non-user space one as well as the impact the optimizations mentioned above have on the overall throughput performance. As we can see, native versus K8s with all optimizations turned on achieves fairly similar performance, while not pinning a VNF to a CPU core and isolating the core from the operating system can drastically reduce this performance in the presence of "noise," e.g. another process/VNF being scheduled on the same core.

7.4.1 High Volume Servers' Software and Hardware Optimization for Packet Processing

7.4.1.1 Data Plane Development Kit [27]

To achieve the best possible data plane performance in terms of throughput, latency and jitter the S/PGW-U data plane functionality is implemented using the Data Plane Development Kit (DPDK). DPDK provides a highly efficient and high performance user mode I/O library bypassing the kernel network stack, thus allowing a very high throughput per CPU Core. As an example from the DPDK Performance Report [28], we see in Figure 7.15 the throughput achieving 50 Gbps line rate at 64 byte packets with an Intel® Xeon® Processor Platinum 8180 (38.5 M Cache, 2.50 GHz) using 2-core, 2-thread, 2-queue per port using the XXV710-DA2 NIC adapter.

Figure 7.16 gives internal details of the S/PGW-U data plane functionality where each individual block of the processing pipeline follows a match/action semantic, with a lookup on tunnel end-point ID, IP address or other identifiers looking for a match and then a specific action depending on the processing block.

To accommodate performance requirements, this data plane can scale-up with the addition of CPU cores associated with the running instance. For example, a software load balancer can front the data plane pipeline, read data from the NIC,

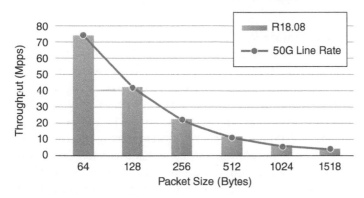

Figure 7.15 DPDK L3fwd performance.

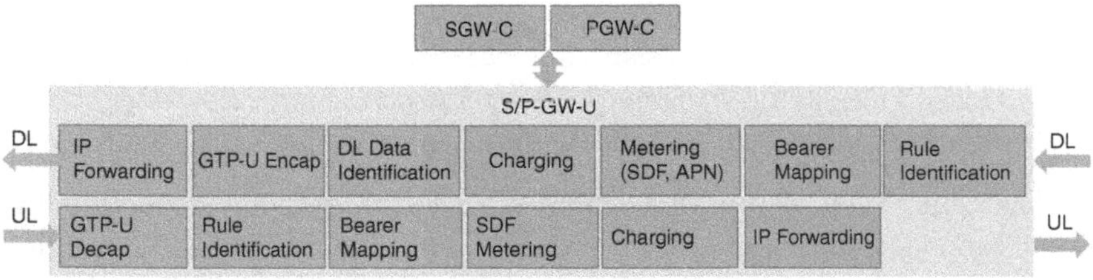

Figure 7.16 Example of OMEC data plane.

and perform flow-based load balancing to the CPU cores. Or the data plane pipeline can take advantage of the NIC's supporting multiple Rx/Tx queues and using Receive Side Scaling (RSS) [29] to assign such queues to CPU Cores. As the incoming traffic on the S1-U interface from the eNB to the S-GW is encapsulated in a GPRS Tunneling Protocol User Plane [30] frame, then in order to have a large number of entries for an efficient RSS, the RSS key should be on the inner IP address in the tunnel, i.e. the user-element IP address instead of the external IP address of the tunnel, the eNB IP address. Latest NIC from Intel supports Dynamic Device Personalization [31], allowing the programmer to setup the RSS keys on any field of the incoming payload, which in this case includes performing the hash on the user-element IP address. This is not an issue on the SGi interface of the P-GW facing the Internet as the traffic is IP-based on this interface. If operating in a virtual environment a NIC supporting SR-IOV [32] offers a single PCIe interface supporting multiple VFs to different VMs or containers.

Having a match/action processing pipeline requires a very efficient lookup mechanism which does not degrade the throughput performance as the number of entries in the lookup table grows to millions.

7.4.1.2 Flow Classification Bottleneck

Flow Classification is a common stage for many network functions, and microservices. A common task for the processing stages shown above is Flow Classification. Flow Classification is mainly given an input packet, some identifiers are extracted (e.g. header information) and using these identifiers a flow table is indexed and a certain action is determined based on the lookup result. For example, determining the output port for a packet, or determining an action (e.g. drop, forward, etc.) for a flow, or an address translation table to determine a tunneling ID to encapsulate the packet with. Since Flow Classification needs to be done at line rate, it is typically a performance bottleneck and many specialized network appliances use ternary content addressable memory and hardware accelerated classifiers to optimize this stage.

The three main performance metrics for a flow table design are:

- Higher lookup rate, as this translates into better throughput and latency.
- Higher insert rate, as this translates into faster updates for the flow table.
- Efficient table utilization, as this translates into supporting more flows in the table.

When running a network function on general purpose servers, Flow Classification is implemented in software, as a result it is crucial to optimize the performance of this stage on general purpose computer hardware (with respect to each of the abovementioned metrics).

7.4.1.3 Cuckoo Hashing for Efficient Table Utilization

As previously mentioned, since each connection in the system has its own state, the number of entries in the system grows with the number of users and devices connected to the network. With 5G, having millions of devices and flows concurrently active on the network, designing a high-performance flow table that can handle millions of flows is a must-have. A Cuckoo hashing flow table (e.g. [33–38]) provides such high performance as it relies on a bucketized hash table which provides excellent lookup performance by guaranteeing that the flow classification can be found in a fixed amount (basically two) of memory accesses, as well as providing very high utilization because it uses multiple hash functions and a table scheme in which the inserted elements can be moved among the tables to make space for new items (typically, for random keys a Cuckoo hashing-based flow table supports inserting flows to almost 95–98% of maximum table capacity). The basic idea of Cuckoo hash-based flow tables is shown in Figure 7.17.

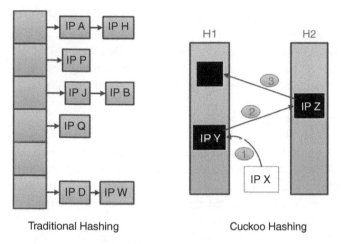

| Traditional Hashing | Cuckoo Hashing |

Figure 7.17 Cuckoo hashing for efficient flow table.

Usually a hash table (<key, value> pair) is used to implement an exact match classifier where the hash key is the flow identifier and a hash function is used to compute an index into an array of buckets or slots, from which the correct value can be found. Ideally, the hash function will assign each key to a unique bucket but this situation is rarely achievable and in practice collisions are unavoidable. The traditional hash table typically uses a single hash function, as shown in Figure 7.17, and since hash collisions are practically unavoidable, the total number of flows inserted is limited (actually, with random keys, and for traditional hashing the first flow that cannot be inserted in the table occurs at a table load of as low as only 7–10%). Cuckoo hashing, on the other hand, ensures a constant lookup time in the worst case. It uses two (or more) hash functions and the key/value pair could be in two (or more) locations, as shown in Figure 7.17.

The data structure used in Cuckoo hashing is a set of "d" hash tables, such that an element x can be placed in tables 1, 2, ..., d in positions $h1(x)$, $h2(x)$, ..., $hd(x)$ given by a set of "d" independent hash functions (typically $d = 2$). In the general case, each position is a bucket and can store up to b elements. For lookup, and given a key X, the key is hashed and the bucket at position $hi(x)$ is accessed and all the b elements (if any) in that bucket are compared with x for a hit or miss. If there is no match, the second bucket location at position $hi + 1(x)$ is looked up, and the process is repeated, until a match is found, or all d locations have been searched. The constant lookup time (at most "d" memory accesses), comes at the cost of a more complex insert operation. To insert an element x, the bucket at position $hi(x)$ is accessed. If it is empty, the new element is inserted there. If not, the next bucket position at $hi + 1(x)$ is tried for any empty slot. If no empty position is found for all d tables, the key x is stored in its primary bucket location $h1(x)$, replacing another key y there. The insertion process is repeated for key Y where all the d locations for it are tried, and if no empty location is found, key Y will replace another key Z from its primary bucket at $h1(Y)$ and the process is repeated for Z and so on until all keys have been placed in a location in the hash table. This insertion procedure is recursive and tries to move elements to accommodate the new element if needed.

7.4.1.4 Intel Resource Director Technology [39]

As the virtual EPC has been disaggregated, separating control plane from the data plane, and the functionality implemented as network functions, this led to the creation of a series of VMs, containers or processes to design and implement the new architecture. When these resources are positioned at run time on the same host, they will contend for the host's shared resources. On a chip multiprocessor, there are shared resources across processes/threads such as the last level cache (LLC), memory bandwidth, interconnect bandwidth or physical I/O devices. The consolidation of various workloads, from control plane to data plane related workloads, or with an application that can be called a "noisy neighbor," will impact shared resources usage and potentially impact performance such as throughput and latency. In general applications, streaming can cause LLC excessive cache line eviction, usually impacting throughput and latency, two critical properties of the SGW and PGW data plane.

Figure 7.18 represents a high level view of Intel Resource Director Technology, which is a set of technologies embedded in the silicon to monitor and enforce shared resources usage. For example, the LLC usage can be monitored through software to identify LLC occupancy on a per-resource basis, e.g. process, thread, VM, and then through software control partitioning of the LLC can be performed to isolate and separate the resources to use a determined and dedicated amount of

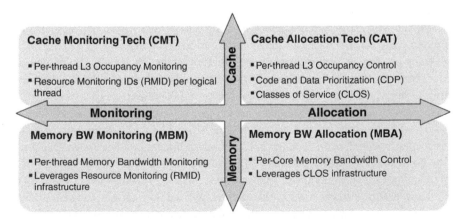

Figure 7.18 Intel Resource Director Technology.

LLC. Similarly, the memory bandwidth usage can be monitored and allocated to appropriate resources.

The academic community [40] has shown the benefits that can be achieved by monitoring shared resources usage and then allocating these shared resources to specific VMs. In an environment with nine VNFs executing simultaneously on isolated CPU cores, without isolation of the LLC shared resource, the LLC contention causes up to ~50% throughput degradation for some of the VNFs, others were minimally impacted due to their workload limited usage of the LLC, and a latency degradation of up to ~30%. When the shared LLC resource was properly allocated and isolated for specific VNFs, the throughput degraded by up to ~2.5% and the latency by up to ~3%, a significant improvement showing the key benefit of these technologies.

In the upcoming 5G network edge environment where we can expect to have most functionality being implemented at VNFs running on high volume servers, use of this technology will be key to guarantee a level of QoS.

7.5   Part III: Software-Defined Disaggregated RAN

7.5.1   RAN Disaggregation

Disaggregation of the RAN aims to deal with the challenges of high total cost of ownership, high energy consumption, better system performance by intelligent and dynamic radio resource management (RRM), as well as rapid, open innovation in different components while ensuring multi-vendor operability.

3GPP has already defined a number of disaggregation options. These are summarized in Figure 7.19.

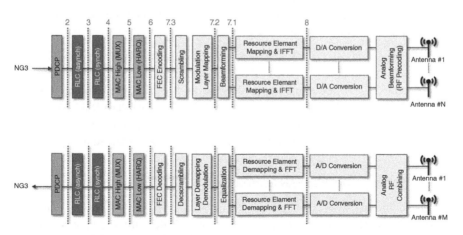

Figure 7.19   3GPP specified RAN disaggregation options.

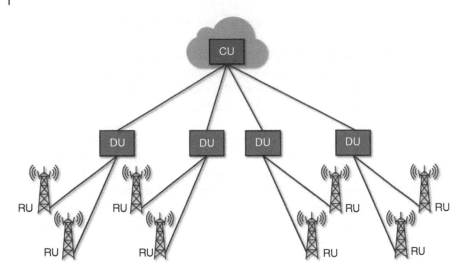

Figure 7.20 RAN disaggregation and distributed deployment.

The disaggregation solution needs to enable a distributed deployment of RAN functions over the coverage area, as illustrated in Figure 7.20, where:

- CU will centralize the "packet processing functions," realize them as virtualized network functions running on commodity hardware, and place them in geographically centralized telco cloud locations.
- DU will realize "baseband processing functions" across cell sites, realize them as virtualized network functions running on commodity hardware, allowing for possible hardware acceleration using, e.g. FPGAs.
- RU will enable geographical coverage using "radio functions" across antenna sites, realized on specialized hardware.

The disaggregation solution needs to be flexible, in that, based on the use case, geography, and operator choice, it should enable the possibility of realizing the base stations as (i) three components (CU, DU, and RU), (ii) two components (CU and DU + RU, and/or CU + DU and RU), or (iii) all in-one (CU + DU + RU).

The current industry trend has focused on an option 2-split between CU and DU and an option 7/8-split between DU and RU. This is illustrated in Figure 7.21.

7.5.2 Software-Defined RAN Control

In the cellular network, RAN provides wide-area wireless connectivity to mobile devices. Then, the fundamental problem it solves is determining how best to use and manage the precious radio resources to achieve this connectivity. On the path

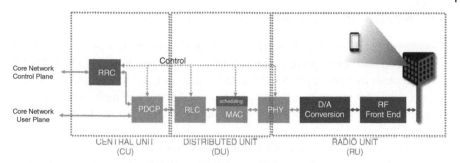

Figure 7.21 Current trend in CU–DU–RU disaggregation.

toward 5G, with network densification, and the availability of different types of spectrum bands, it is increasingly a more difficult task to allocate radio resources, implement handovers, manage interference, balance load between cells, etc. The current RRM control that is distributed across the RAN nodes (base stations) is suboptimal in achieving the conduct of these functions in an optimal way. Thus, it is necessary to bring software-defined controllability to RAN to increase system performance.

The fundamental goal is to allow for programmatic control of RRM functions. This can be achieved by decoupling the associated intelligence from the underlying hardware and stack software. Within the RAN, the RRM functions (Figure 7.21) reside in the CU (in the RRC), and in the RU (in the MAC). Then, the decoupling of the RRM functions from the underlying stack software effectively means:

- Further disaggregating the CU into user and control planes: CU-U and CU-C, where CU-C contains all RRC-side RRM functions.
- Clearly defining open interfaces between CU-C and CU-U, as well as between CU-C and DU.
- Realizing the CU-C as an SDN-controller.
  Explicitly defining a "Radio Network Information Base (R-NIB)," using:
  – RAN nodes: CUs, DUs, RUs, mobile devices (UE, IoT, etc).
  – RAN links: all links between nodes that support data and control traffic.
  – Node and link attributes: static, slow-varying and fast-varying parameters that collectively define the nodes and the links.
- Maintenance and exposure of the R-NIB using an open interface to RRC-side RRM functions that are realized as SDN applications.
- Allowing for programmatic configuration of the MAC-side RRM functions using open interfaces.

The software-defined control of the RAN will allow for:

- *Democratization of the innovation within the RAN*: The control plane–user plane separation and open, clearly defined interfaces between the disaggregated RAN components as well as between the RAN control and associated control applications, allow for innovative, third-party control solutions to be rapidly deployed regardless of which vendors have provided the underlying hardware and software solution.
- *Holistic control of RRM*: Software-defined control of RAN will allow for logically centralized (within limited-geography) control of RRM. Then, for a given active user, using innovative control applications, operators are empowered to conduct dynamic selection of any radio beam within reach across all network technologies, antenna points, and sites using a global view that minimizes interference, and thus maximizes observed user QoE. This can be achieved by applying carrier aggregation, dual connectivity, coordinated multi-point transmission as well as selection of multiple-input multiple-output and beamforming schemes using a global view of the wireless network.
- *Use-case-based management of the RAN*: Software-defined RAN control will allow for the integration of performance-based decisions with policy-based constraints, with such constraints to be dynamically set, based on use cases, geographies, or operator decisions.

The architecture that allows for software-defined RAN control is illustrated in Figure 7.22. In the figure, we note that the RRC is disaggregated into two distinct functional components: Core Network Control Plane Forwarding, which is responsible from forwarding control plane traffic between the UEs and the mobile core network, and RAN Near Real-Time Software-Defined Control, which is responsible from conducting all RRC-level RRM functions, as

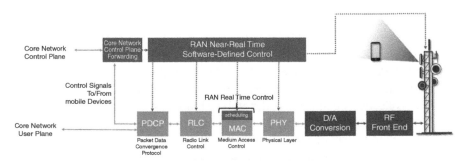

Figure 7.22 RAN near real-time and real-time software defined control.

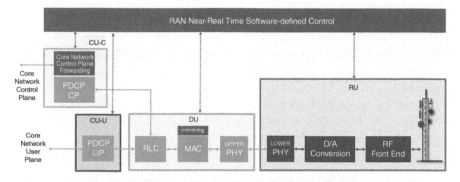

Figure 7.23 Disaggregated and software defined controllable RAN architecture.

well as SON control applications in a programmable way. Following our SDN-based terminology, we also describe the MAC-level RRM function, namely the scheduler, as the RAN Real-Time Control in the figure. While the Real-Time Control is not disaggregated and logically centralized in this architecture, it is possible for this component to be programmatically configured by the Near Real-Time Controller.

Combining disaggregation and software-defined control results in a RAN architecture is illustrated in Figure 7.23. As observed, the disaggregated, and software-defined controllable RAN is composed of RAN Near Real-Time Software-Defined Controller, CU-C, CU-U, DU, and RU components. While RU and DU are distributed across the geography across antenna sites and cell sites, respectively, the near real-time controller, CU-C and CU-U can be instantiated as virtualized network functions at edge clouds.

7.6 Part IV: White-Box Solutions for Compute, Storage, Access, and Networking

The rise of white box for compute, storage, and networking is shaping how modern networks will take advantage of cloud technologies at the edge of the network to meet the demand of end devices for better experiences. High volume servers, switches built using merchant silicon, and appliances can now perform what used to be dedicated to special hardware and ASICs such as signal processing, packet processing, and adopt open source software available for the "white box."

Today most RAN, switches and optical deployments are hardwired and static, inhibiting dynamic services on demand and flexibility, leading to time-sensitive truck roles, higher costs, and loss of revenue opportunities for all. For example, white box deployed RAN for emerging 5G networks will enable real time control over microservices that were previously at the mercy of a specific implementation approach until a new "G" generation arrives:

- Open Disaggregated Interoperable RAN and CORE network layers that can be upgraded on a software development time scale versus a hardware development time scale.
- Standard-based API from each component that can enable innovation.
- Disaggregated and software controlled radios that span into split architecture options between, e.g. 5G RUs, 5G DUs, 5G CUs.
- Run time SW plug-ins to meet the end user's device behaviors and policies.

In general, the "white box" term has been applied to no-name computers. The same ODMs (original design manufacturers) that produced them are getting into the game of manufacturing white-box platforms that meet certain standards such as OCP (Open Compute Platform; https://www.opencompute.org). White box platforms look just like any other purposeful hardware such as switch, chassis, or high volume servers. Furthermore, white boxes promise cost savings and greater flexibility over traditional purposeful hardware appliances.

Google G-Scale Network [14] has shown some of the benefits of using white box platforms built using merchant switching silicon to enable new functionalities such as disaggregation of control from data planes, i.e. SDN, and taking advantage of new services. White box platforms, e.g. switches, come with their operating systems, usually Linux based and ideally hardware agnostic, and specific software stacks to manage and set up their control and data plane. For example, in the case of a switch or white box radio or IP equipment, the software needs to seamlessly integrate with existing L2/L3 topologies and support a basic set of features. Beyond the separation of control from data planes and availability of open source software, white box platforms are more valuable if they interact with SDN controllers. Some of these white box products, e.g. switches, come with software control which interfaces to multiple controllers. By controllers we mean software defined everything where white box platforms can be configured or imaged to be a network appliance, a switch, or even a RAN. Beyond this, there could be new capabilities delivered as a result of "opening up" the network functions, e.g. switch, appliances, etc. as well as being able to use many new open source tools that can be used, that is the best foundation for reaping the benefits of network functionality implemented on white box. The advantage of this approach is that network functions, e.g. switches or others, can be customized to meet an organization's specific business and networking needs. Due to their flexibility, white box platforms can support a range of open source tools to operate, control, and manage networks with more solutions from various consortiums coming to markets. To name a few, the ONF Central Office Re-architected as Datacenter [4] defining multi-access edge network running on high volume servers, the ONAP [5], a platform for real-time, policy-driven orchestration and automation of physical or VNFs providing a complete lifecycle

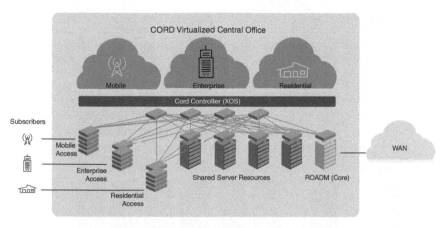

Figure 7.24 CORD high-level architecture.

management to on-board and manage functionality, Akraino [41], an integration platform for the edge of the network, and the Open Platform for NFV [42], an integration platform facilitating the development and validation of VNFs.

For example, ONF's open source CORD edge cloud hardware reference model [43, 44] illustrated in Figure 7.24 has commodity off-the-shelf servers (COTS) inter-connected by white box software defined programmable switches and specialized or white box access hardware to connect the end-users (residential, commercial, or mobile). ONF recommends OCP-certified hardware but this is not an essential requirement.

CORD integrates the relevant open source components packaged as a standalone POD that is easy to configure, build, deploy, and operate. Various configurations of CORD are being deployed in production networks by major carriers around the world, making it an ideal experimental platform with a compelling tech transfer story. CORD's software reference model is shown in Figure 7.25.

CORD arranges the software defined programmable switches in a leaf-spine fabric, enabling a highly scalable platform. It orchestrates the servers with Kubernetes. It is also possible to engage OpenStack, or the combination of the two for this purpose. The software defined control of both the switching fabric and the access devices are conducted by ONF's ONOS controller and associated ONOS applications. In Figure 7.25, two access technologies are illustrated: mobile and broadband. For mobile access, ONOS controls the disaggregated, or all in one base stations using the near real-time RAN control application. For broadband access, white box realizations of optical line terminations (OLTs) are turned into SDN controllable forwarders via ONF's virtual OLT hardware abstraction (VOLTHA) platform. Subsequently, a virtual OLT (vOLT) application running on ONOS performs the associated control. XOS is CORD's services controller and it integrates the disaggregated components into a self-contained system. It defines a set of abstractions that govern how all the individual components

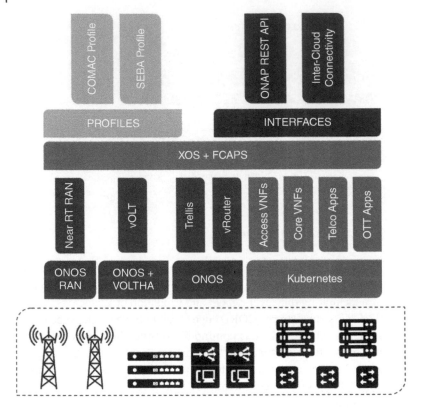

Figure 7.25 CORD software reference model.

interconnect to form a service mesh, how individual subscribers acquire isolated service chains across the service mesh, and how operators monitor, provision, and configure CORD. For monitoring, XOS engages open source platforms such as Prometheus, Grafana, Kibana, etc. Users wanting to differentiate their offering could replace specific module or work with the open source community to improve functionality or performance of specific components or the system as a whole.

7.7 Part V: Edge Cloud Deployment Options

There is no single physical location that can be defined as the network edge. In fact, as illustrated in Figure 7.26, multiple edge locations in the network exist:

- *Device edge*: Many of the end user devices are furnished with significant compute, storage and network resources. This effectively enables the zero-hop device edge, where potentially some workloads may run, especially those with

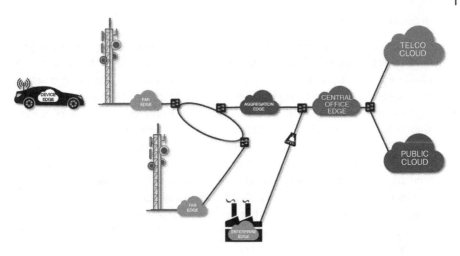

Figure 7.26 Edge cloud locations.

the most stringent latency requirements. The device edge needs a remote control and management platform. This will be located at either another edge cloud location, or at the central (telco or public) cloud.

- *Far edge*: An edge cloud may exist at every base station site, resulting in a cell-specific edge cloud platform. The far edge has physical limitations in real estate and power, making it resource constrained. At the same time, the far edge is the closest network-side edge location, potentially providing lowest latencies and higher throughputs than other network-side edges. The constrained resources coupled with the potential for high performance mean that intelligent workload placement orchestration overseeing efficient utilization of far edge resources is critically important.
- *Aggregation edge*: An edge cloud may also exist at the base station aggregation sites. Relative to the far edge, the aggregation edge may have more resources for pooling, allowing workloads serving geographies covered by multiple base stations to be hosted here.
- *Central office edge*: A central office (CO), owned by an operator, is usually located in the center of a town, host switching and routing systems that allow for local traffic switching as well as routing to other central offices, or external networks for wireline, fiber and mobile access [45]. Up until recently, systems in the COs have been very hardware centric. Today, increasingly, we observe such systems being replaced by VNF workloads that resemble those found in cloud or datacenters. Therefore, a CO is the first natural operator-owned edge cloud location.

In addition to these locations, an additional location exists:

- *Enterprise edge*: In parallel with operators' efforts on defining and deploying edge clouds, cloud providers are also making a play. For them, the natural edge cloud location is the enterprise datacenter. The focus here is to provide machine learning and inference-based intelligence to enterprises' IoT solutions by bringing models that used to run in the cloud down to the edge. Unlike operators' approaches, the last mile connectivity between the enterprise edge and the end user and IoT devices is abstracted out from the cloud providers' edge, provided that this connectivity does not bring substantial performance degradation. Similarly, the enterprise edge is agnostic to the backhaul technology that enables its connectivity to the public cloud. This connection can be fiber, wireline, or mobile. One needs to ensure that this connection satisfies the requirements of the specific machine learning and inference intelligence that will be distributed between the edge and the cloud.

Figure 7.26 provides an overview of all of the above described edge locations.

In the distributed network cloud paradigm, operators' disaggregated VNF workloads will be distributed across the central telco cloud and edge cloud locations so that performance requirements for different business verticals and use cases are met while any potential efficiency is leveraged via resource pooling. These workloads include the virtualized components of the RAN (DU, CU-U, and CU-C), software-defined controller for the RAN (SD-RAN), core network user plane function (UPF), core network control plane functions (CPFs) as well as operator provided applications such as voice, video streaming, and AR/VR. Similarly, workloads for machine learning and inference-based intelligent IoT operations (Inference, ML) as well as any enterprise and/or OTT applications and services (App) will be distributed across the public cloud, optionally the telco cloud, as well as the edge cloud locations. A sample workload distribution across the distributed network cloud is illustrated in Figure 7.27.

A more involved synergy between the telco network cloud and the public cloud is also possible. The disaggregation of the access and connectivity VNFs allows a deployment of components where data plane components are hosted at operators' central and edge cloud locations, and the control plane components are hosted at a public Cloud Service Provider's cloud, as shown in Figure 7.28.

A deployment example has been demonstrated by ONF at the Mobile World Congress in 2019. In this demonstration, an edge cloud located at ONF's booth in Barcelona was connected to two central Telco Cloud locations, one inside Deutsche Telekom's datacenter in Warsaw, and another inside Turk Telekom subsidiary Argela's datacenter in Istanbul. The edge cloud platform ran CORD to control and manage COTS, white box switches as well as programmable access equipment. The edge cloud in question provided a shared platform for two specific access

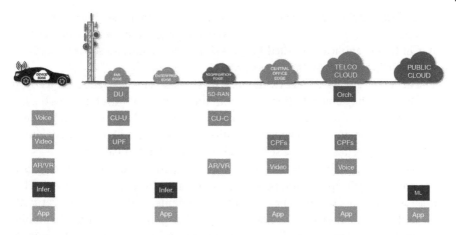

Figure 7.27 Sample distribution of VNF and application workloads across a distributed network cloud.

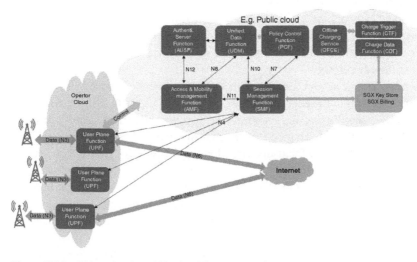

Figure 7.28 Telco cloud–public cloud inter-operation.

technologies, broadband and cellular. As such, the OLT was realized as a white box solution with a VOLTHA enabling Openflow-based programmability to broadband access using a vOLT application running on ONOS. The broadband service subscriber control was conducted by a simplified virtual broadband network gateway that was realized as a VNF hosted at the edge cloud.

For cellular access, a disaggregated, commercially available small cell that allowed for programmatic control was utilized. The CU was realized in a microservices architecture using three distinct containerized, inter-connected components. The SD-RAN control allowed for programmatic network slicing using an ONOS application called ProgRAN. Further details on software defined network slicing is given in the Section 7.8.

The cellular network core network was distributed across the edge and central telco clouds in the demonstration, with containerized workloads for UPFs hosted at the edge, and all containerized CPFs hosted at the central telco cloud in Warsaw. The global end-to-end network orchestration was conducted by the open source ONAP which was hosted at the second central telco cloud in Istanbul.

A CDN-based streaming video service was included in the demonstration as a representative OTT application that could share a multi-tenant edge cloud. This service was developed using open source components, with NGINX server with local content cache hosted at the edge cloud providing streaming service to mobile handsets with very low pre-roll latency and uninterrupted streaming experience. A remote content store was developed using VLC and remote streaming of this content to NGINX server was enabled by Wowza. Both VLC and Wowza were hosted at the central Telco cloud in Warsaw.

An optical network unit, representing a broadband subscriber location, was connected to the OLT hosted inside the edge cloud. Similarly, a couple of mobile handsets were connected to the base station with the virtualized components hosted at the edge cloud.

Two network slices were designed on ONAP, and the corresponding profiles were pushed down to the edge cloud for execution. CORD's XOS, upon receipt of these profiles, coordinated the creation and execution of these slices using the ProgRAN application running on ONOS. One slice was for users streaming video content and another was for users downloading a file with best effort Service Level Agreement.

A high-level diagram of the ONF demonstration is given in Figure 7.29. As shown in the diagram, the demonstration enabled isolation between the user plane and control/management plane traffic by using two distinct network interfaces on Kubernetes. While the user plane enjoyed acceleration via an SR-IOV interface, the control/management plane utilized Calico. It should be noted here that due to the nature of the cellular network realization, this meant that the core network user plane container required dual connectivity, one for the Calico network, and another for the SR-IOV network. This was enabled by making use of Multus.

7.8 Part VI: Edge Cloud and Network Slicing

Network slicing is a mechanism to create, and dynamically manage functionally discrete virtualized networks over a common 5G infrastructure, one for each use case/tenant. It realizes the expected benefits of virtualization by providing a

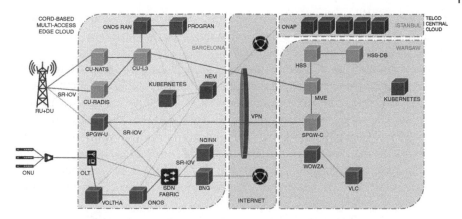

Figure 7.29 ONF's distributed network cloud architecture demonstrated at Mobile World Congress 2019.

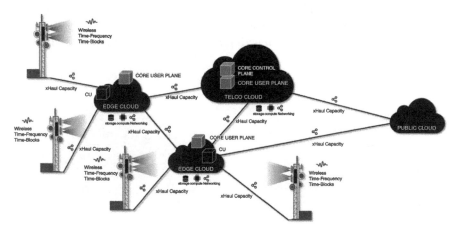

Figure 7.30 5G network resources and capabilities offered as-a-service to use cases and tenants with network slicing.

model for operationalizing virtualized, and programmable cloud-based infrastructure. Then, leveraging disaggregation, NFV and SDN, network slicing is a mobile networking platform using scalable and elastic access, network and network resources provided "as-a-service" to use cases and tenants in a given geography using 3GPP-standardized technologies. As illustrated in Figure 7.30, the network resources comprise storage, compute, and networking resources distributed across central and edge cloud locations, and service resources include virtualized network functions that are instantiated at various locations in the distributed 5G network. If network slicing was to include slicing of the RAN as well, then we would need to add the service resources of disaggregated RAN virtualized network functions as well as access resources of the time-frequency resource blocks to the list of resources that are to be provided to use cases and tenants in a scalable and elastic manner.

For a given use case or tenant, a network slice is designed by specifying slice attributes that include latency, throughput, reliability, mobility, geography, security, analytics, and cost profile parameters. These attributes are then translated into composable access, network, and service resources that are selected in optimally selected geographical locations, including the edge cloud. A key component of network slicing is automated composition of slice specific service chains and their life-cycle management. The designed and composed service chains with associated elastically and scalable allocated resources for the slices are then delivered for the specific use cases or tenants.

An example is illustrated in Figure 7.31, where two network slices are constructed for distinct use cases. The first slice has distinct RAN CP (CP1) chained to a core user plane (UPF1) service, both realized at the edge, enabling a local break-out. A component of a tenant specific application is also instantiated at the edge and is attached to the egress of this service chain. The user plane part of the chain is completed by connecting the remainder of the tenant specific application that resides in the public cloud. The second slice also has a slice specific CP (CP2) and core network user plane (UPF2) chained together.

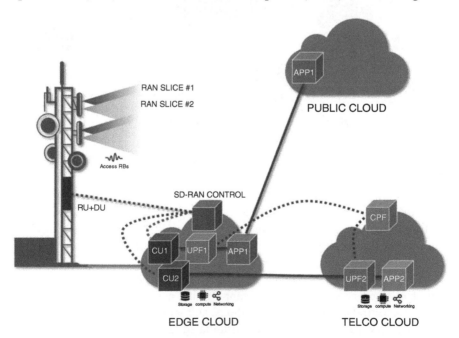

Figure 7.31 Two sample network slices and associated service chains.

However, this time, the UPF workload is instantiated at the telco cloud. A tenant specific application is also instantiated at the telco cloud for the second slice. In the example, both slices share the near real time SD-RAN controller VNF and as well as the core network CPF VNFs. The time-frequency mobile resource blocks are divided in an elastic and scalable way among the two network slices.

Network slicing enables the last component of the network transformation, namely, cloudification. It enables multiple use cases and tenants to share the access, network, and service resources of a common, distributed infrastructure, thereby catalyzing new revenue opportunities for the operators.

Due to its inherent nature, the edge cloud is resource limited. In a given geography, so is the RAN. Therefore, workload placement and automation of resources allocation are crucial components of network slicing.

7.9 Summary

With 5G, a transformation of the network architecture is underway, paving the way to a distributed network cloud. In this new paradigm, edge cloud plays an essential role, as a location providing traditional cloud resources of compute, storage, and networking, as well as network cloud specific resources, such as spectrum in close proximity to the end users. This in turn, allows operators to provide optimized access and connectivity options to different use cases and business verticals.

The edge cloud is where access meets the cloud. As such, it would be unwise to develop an edge cloud architecture that concentrates on a solution just for access or just for cloud. In this capacity, the overall architecture should allow for:

- Hosting of operators' VNFs by providing all necessary scaling, elasticity, and networking requirements.
- Programmatic creation and maintenance of service chains that may transcend a single edge cloud, and potentially include multiple edges, and central clouds, based on recipes provided by global orchestration.
- Hosting of latency, bandwidth, and traffic localization sensitive end user and OTT applications, functions and microservices.
- Federated network slicing that includes service chains of operator VNFs and end-user/OTT applications.

Acknowledgments

The authors would like to thank Tom Tofigh of QCT and Sameh Gobriel of Intel for their help in writing this chapter.

References

1 Wikipedia (2019). Software-defined networking. https://en.wikipedia.org/wiki/Software-defined_networking (accessed 9 October 2019).

2 ETSI (2013). Network Function Virtualization (NFV). http://portal.etsi.org/portal/server.pt/community/NFV/367 (accessed 9 October 2019).

3 ONF (2018). Open Networking Foundation. https://www.opennetworking.org (accessed 9 October 2019).

4 ONF (2018). CORD. https://www.opennetworking.org/cord (accessed 9 October 2019).

5 ONAP (2018). Open Network Automation Platform. https://www.onap.org (accessed 9 October 2019).

6 Telecom Infra Project (n.d.). OpenRAN. https://telecominfraproject.com/openran (accessed 9 October 2019).

7 O-RAN (2018). O-RAN Alliance. https://www.o-ran.org (accessed 9 October 2019).

8 Telecom Infra Project (n.d.). OpenRAN 5G NR. https://5gnr.telecominfraproject.com (accessed 9 October 2019).

9 Telecom Infra Project (2019). Open RAN 5G NR Base Station. https://5gnr.telecominfraproject.com/wp-content/uploads/sites/27/PG-Charter-5G-NR-Base-station.pdf (accessed 9 October 2019).

10 Vogels, W. (2017). Unlocking the Value of Device Data with AWS Greengrass. https://www.allthingsdistributed.com/2017/06/unlocking-value-device-data-aws-greengrass.html (accessed 9 October 2019).

11 O-RAN Alliance (2018). O-RAN: Towards an Open and Smart RAN. https://static1.squarespace.com/static/5ad774cce74940d7115044b0/t/5bc79b371905f4197055e8c6/1539808057078/O-RAN+WP+FInal+181017.pdf (accessed 9 October 2019).

12 OpenFlow (2011). OpenFlow. https://openflow.stanford.edu (accessed 10 October 2019).

13 P4 (2015). P4 Language Consortium. https://p4.org (accessed 10 October 2019).

14 Koley, B. (2014). Software Defined Networking at Scale. https://static.googleusercontent.com/media/research.google.com/en//pubs/archive/42948.pdf (accessed 10 October 2019).

15 P4 Language Consortium (2017). $P4_{16}$ Language Specification. https://p4lang.github.io/p4-spec/docs/P4-16-v1.0.0-spec.html (accessed 10 October 2019).

16 P4 Language Consortium (2017). P_4Runtime. https://p4.org/p4-runtime (accessed 10 october 2019).

17 Schmitt, P., Landais, B., and Young, F. (2017). Control and User Plane Separation of EPC nodes (CUPS). http://www.3gpp.org/cups (accessed 10 October 2019).

18 Rajan, A.S., Gobriel, S., Maciocco, C. et al. (2015). Understanding the bottlenecks in virtualizing cellular core network functions. https://ieeexplore.ieee.org/document/7114735 (accessed 10 October 2019).

19 Sunay, O., Maciocco, C., Paul, M. et al. (2019). OMEC. https://www.opennetworking.org/omec (accessed 10 October 2019).

20 OpenAirInterface (OAI) (2019). OpenAirInterface. https://www.openairinterface.org (accessed 10 October 2019).

21 Facebook Connectivity (2018). Magma. https://connectivity.fb.com/magma (accessed 10 October 2019).

22 Cloud Native Computing Foundation (2018). kubernetes. https://kubernetes.io (accessed 10 October 2019).

23 Intel/multus-cni (2018). Multus-CNI. https://github.com/Intel-Corp/multus-cni (accessed 10 October 2019).

24 HashiCorp (2017). Consul. www.consul.io (accessed 10 October 2019).

25 Saur, K. and Edupuganti, S. (2018). Migrating the Next Generation Mobile Core towards 5G with Kubernetes. Container Days, Hamburg.

26 Intel Corporation (2017). Enhanced Platform Awareness in Kubernetes. https://builders.intel.com/docs/networkbuilders/enhanced-platform-awareness-feature-brief.pdf. Code at: https://github.com/Intel-Corp/sriov-cni (accessed 10 October 2019).

27 DPDK (2014). Data Plane Development Kit. https://www.dpdk.org (accessed 10 October 2019).

28 Intel (2018). DPDK Intel NIC Performance Report Release 18.08. http://fast.dpdk.org/doc/perf/DPDK_18_08_Intel_NIC_performance_report.pdf (accessed 10 October 2019).

29 Microsoft (2017). Introduction to Receive Side Scaling. https://docs.microsoft.com/en-us/windows-hardware/drivers/network/introduction-to-receive-side-scaling (accessed 10 October 2019).

30 ETSI (2017). ETSI TS 129 281. https://www.etsi.org/deliver/etsi_ts/129200_129299/129281/08.00.00_60/ts_129281v080000p.pdf (accessed 10 October 2019).

31 Intel (2018). Dynamic Device Personalization for Intel Ethernet 700 Series. https://software.intel.com/en-us/articles/dynamic-device-personalization-for-intel-ethernet-700-series (accessed 10 October 2019).

32 PCI SIG (1994). Specifications. https://pcisig.com/specifications/iov (accessed 10 October 2019)

33 Zhou, D., Fan, B., Lim, H. et al. (2013). Scalable, high performance Ethernet forwarding with CuckooSwitch. *ACM Conference on Emerging Networking Experiments and technologies (CoNEXT)*, Santa Barbara, CA.

34 Scouarnec, N. (2018). Cuckoo++ hash tables: high-performance hash tables for networking apps. *ACM/IEEE Symposium on Architectures for Networking and Communications Systems*.

35 Breslow, A.D. (2016). Horton tables: fast hash tables for in-memory data-intensive computing. *USENIX Annual Technical Conference*, Denver, CO.

36 Dietzfelbinger, M. (2005). Balanced allocation and dictionaries with tightly packed constant size bins. *ICALP 2005*.

37 Wang, Y. (2017).Optimizing Open vSwitch to support millions of flows. *GLOBECOM 2017*.

38 Wang, Y. (2018). Hash table design and optimization for software virtual switches. *KBNets 2018*.

39 Intel (n.d.). Intel Resource Director Technology. http://www.intel.com/content/www/us/en/architecture-and-technology/resource-director-technology.html (accessed 10 October 2019).

40 Tootoonchian, A., Panda, V., Lan, C. et al. (2018). ResQ: Enabling SLOs in Network Function Virtualization. https://www.usenix.org/system/files/conference/nsdi18/nsdi18-tootoonchian.pdf (accessed 10 October 2019).

41 Linux Foundation Edge (2018). Akraino. https://www.akraino.org (accessed 10 October 2019).

42 Linux Foundation (2016). OPNFV. https://www.opnfv.org (accessed 10 October 2019).

43 Peterson, L., Al-Shabibi, A., Anshutz, T. et al. (2016). Central office re-architected as a datacenter. *IEEE Communications Magazine* 54 (10): 96–101.

44 Peterson, L., Anderson, T., Katti, S. et al. (2019). Democratizing the network edge. *ACM SIGCOMM Computer Communication Review* 49 (2): 31–36.

45 TIA (2018). Central Office Evaluation Strategy for Deployment of Information and Communications Technology (ICT) Equipment. Position paper.

Part V

5G Verticals – Key Vertical Applications

8

Connected Aerials

Feng Xue, Shu-ping Yeh, Jingwen Bai, and Shilpa Talwar

Intel Corporation, Santa Clara, CA, USA

Abstract

To operate drones safely and to accomplish various missions, wireless technology is a critical component in control and command. This chapter discusses basic requirements and challenges for supporting drones over a cellular network. It provides a quick summary of drone regulations, which serves as a good indication of challenges in supporting drones and innovation in communications. The chapter reviews 3GPP Release 15 drone study and WIs – the first on drones in 3GPP. It provides a 5G drone support deep dive discussing challenges in supporting various drone uses, solutions from new technologies in 5G, and new studies needed for further 5G innovation toward future drone usages. Many drone applications, such as surveying, require 3D positioning support. Since cellular networks have always focused on serving terrestrial users and mostly use down-tilted base station antennas, aerial coverage and link qualities need to be carefully studied and optimized for drone support.

Keywords *aerial communication; aerial coverage; cellular network; drone regulations; drone usages; terrestrial users*

8.1 Introduction

In recent years, aerial vehicles (AVs) such as UAVs (unmanned aerial vehicles, also known as drones) have experienced exponential growth, and many innovative usage scenarios are being considered. For example, drones can be used for surveying, search and rescue, maintenance and repair, safety, communications,

5G Verticals: Customizing Applications, Technologies and Deployment Techniques, First Edition.
Edited by Rath Vannithamby and Anthony C.K. Soong.

image capture, surveillance, etc. These non-conventional usages are useful for many industries such as energy (oil/gas/wind/solar), insurance, telecom, agriculture, shipping, wildlife and environment research, sports, and so on. In many cases, drones could fundamentally transform the industry. Even more futuristic, UAVs are being considered as air taxis to transport humans [1].

To operate drones safely and to accomplish various missions, wireless technology is a critical component in control and command (C2). Currently, the control is typically based on RC (remote control) channel, WiFi and/or Bluetooth, which have limited data rate and transmission range. Since most countries allow only visual-line-of-sight operation of drones now, they may satisfy the immediate needs. However, with so many emerging drone applications and in anticipating future new regulations, the cellular system has been under investigation to support increased requirements on reliability, large scale operation, and high data rate. This is because there are many benefits by utilizing a cellular network:

- With licensed spectrum and well controlled infrastructure, a cellular network can provide dedicated resources for QoS (quality of service) guarantee, especially for reliability and high data rate. This is in great contrast to WiFi/Bluetooth solutions which are on a shared spectrum.
- A cellular network is widely deployed and well connected via fast speed backhaul links. It not only provides data services but also provides services such as positioning, identification, etc. *over a large area.* These services with reliability make it a good candidate as an infrastructure for supporting large-scale and dense operations; it is also a good candidate for autonomous non-visual line-of-sight operations.
- Besides WiFi/RC/Bluetooth-based solutions, a cellular network provides further redundancy to increase reliability which is much needed for the safe operation of drones.

Given the above considerations and huge business potentials, both academia and industry have taken efforts to understand the support of drones over 4G and 5G systems. As one of the important 5G vertical usages, system requirements related to drone cases have been incorporated into 5G system design requirements [2]. Performance impact and challenges due to drones have been studied in 3GPP Release 15 WIs [3], resulting in the introduction of a set of features for continuous 4G/5G evolution. With many new technologies introduced, 5G cellular will enhance the support of drones with more advanced interfaces such as ultra-reliable and low-latency communications (URLLC). Millimeter wave (mmWave) may satisfy the need of large capacity for transferring a large amount of data from drone to ground (e.g. 4K/8K video). Interestingly, 5G development will also benefit from the continuous development of new drone usages. For example, drones may be employed to serve as flying cells for temporary coverage enhancement.

The chapter is organized in the following way. Section 8.2 discusses basic requirements and challenges for supporting drones over a cellular network. Section 8.3 provides a quick summary of current drone regulations, which serves as a good indication of challenges in supporting drones and innovation in communications. Section 8.4 reviews related R&D efforts, focusing on communications from both academia and industry. Section 8.5 reviews 3GPP Release 15 drone study and WIs – the first on drones in 3GPP. Section 8.6 provides a 5G drone support deep dive discussing challenges in supporting various drone uses, solutions from new technologies in 5G, and new studies needed for further 5G innovation toward future drone usages.

For ease of reading, the acronyms used have been summarized at the end of the chapter. Note in many places the words UAV, drone, and aerial are interchangeable.

8.2 General Requirements and Challenges for Supporting UAVs over a Cellular Network

AVs, even though in a certain sense they can be considered as a subset of autonomous vehicles, have several unique properties compared with traditional terrestrial users and other verticals. This makes them a unique category worthy of special consideration in 5G.

First of all, the drone activity region could be far above ground level. In current regulations, drones are allowed to fly 100–300 m or even higher depending on the country's regulations. Since cellular networks have always focused on serving terrestrial users and mostly use down-tilted base station antennas, aerial coverage and link qualities need to be carefully studied and optimized for drone support. As measurements have shown, drones cause and receive excessive interference due to slower signal attenuation in the air compared with on the ground. In addition, as illustrated in Figure 8.1, antenna side lobes complicate the situation further [3] due to their fast and irregular fluctuations.

AVs can continuously move at fast speed in the sky over a large area. Although it might seem that challenges in mobility support may have some similarity to ground vehicle scenarios, there are some differences. First, the drone movement is 3D. Drones could be moving in any arbitrary fashion without the constraint to be along predefined routes like ground vehicles. Unique properties of the aerial channel also play a role. Due to nicer propagation properties in the air, a drone sees a larger number of base stations than a car on the ground as verified by both measurements and simulations. As it is supported by the side lobes, signal qualities from many far away base stations are comparable with local ones. Thus a fast-moving drone can experience communication links with rapid fluctuation in quality. Thus, careful design of drone mobility support is needed.

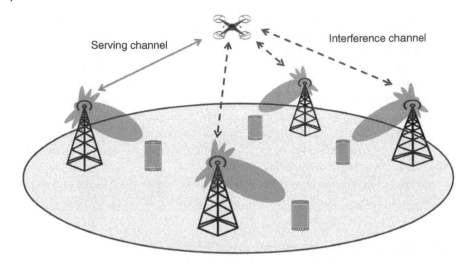

Figure 8.1 Interference situations in the air for drones.

Many drone applications, such as surveying, require 3D positioning support. Accurate positioning is particularly important for safety-critical procedures, such as fly planning and collision avoidance. Even though positioning can be provided by GPS in an outdoor case, redundancy is needed as GPS could be subject to blockage and/or multipath in many scenarios.

Special traffic pattern in drone usages may also demand enhanced wireless support. Contrary to the traditional view that the cellular network is dominated by downlink traffic, uplink data traffic is dominant in many drone use cases when drones need to send high quality video streams to users or data centers. Communication support for edge-computing needs to be addressed as well. For example, in the case of real-time surveillance, a drone's video stream needs to be processed for real-time reasoning. This involves both data analytics and machine learning at the edge.

Public safety is another key concern regarding drones. Regulation agencies have been considering possible solutions for ensuring public safety. 4G/5G cellular systems can enhance identification, localization, geo-fencing, etc. As more advanced operations such as beyond-visual line-of-sight, drone swarm, and air taxi come into the picture, requirements on extra reliability, throughput and computing support from the cellular systems are significantly increased. More details on these emerging operations will be discussed later.

Summarizing these observations, the following are several basic challenges to be considered and enhanced for supporting drones on cellular networks:

- Interference mitigation
- Mobility support enhancement
- Latency and reliability

- 3D Positioning
- Enhanced throughput in uplink
- Edge computing support
- Safety

8.3 Summary on Current Drone Regulations

Regarding communication support, most wireless technologies such as PAN and WAN have been designed and regulated based on terrestrial uses. Spectrum allocations and regulations are under investigation for aerial usages.

Regarding drone operation regulation, there have been early developments. In the United States, the Federal Aviation Agency (FAA) established a set of rules on small drones between 0.5 and 55 lb in Part 107 [4] in 2016. It sets very restrictive constraints on how one can operate drones and is representative of regulations across the globe. The following are a few samples from Part 107:

- Non-visual-line-of-sight operation is *not* allowed.
- One pilot cannot operate more than one drone.
- Drone cannot fly over 400 ft. above ground or a building.
- Drone speed should be less than 100 miles per hour.

In other countries such as Canada, China, Japan, and the EU, similar rules have been implemented (e.g. China [5]). All are in an early stage as drone operation scenarios are still undergoing fast changes and growth. Safety has been a main concern. In the meantime, various programs have been set up to encourage drone innovations in the future.

8.4 Review of Aerial Communication R&D Activities in General

8.4.1 R&D Activities from Industry and Government Agencies

The communication link for current drones is typically based on RC, WiFi, Bluetooth, and other technologies on the ISM-Band. Their limitations are obvious in terms of capacity, reliability, distance, and safety. Efforts have been taken by industry and government agencies to enhance the communication links via several methods.

Qualcomm established a flight test site around its headquarters to study how 4G/5G can help drones [6], with results summarized in [7]. In [8, 9], it claims that "5G will enable wide scale deployments of mission-critical drone use cases." DJI and several startups introduced proprietary solutions to enhance video streaming quality and control distance. AT&T considered improving Long-Term Evolution (LTE) coverage for both ground users and drones at packed venues such as sport activities or in a disaster zone [10]. Verizon kicked off an Airborne LTE Operations (ALO) initiative in October 2016 for innovation on in-flight wireless connectivity.

GSMA [11] and CTIA [12] have published white papers with partner companies claiming and promoting improved drone support over cellular networks. Based on field study, the Civil Aviation Administration of China [13] clarified a set of requirements on operation of drones and claimed cellular infrastructure to be a good candidate for safe drone operation. 3GPP completed an aerial specific Study Item (SI) and Work Item (WI) in RAN1/RAN2 in 2017–2018 [14], and also started working on drone ID at core network side [15].

In existing cellular networks, many spectrum regulations/allocations are based on terrestrial considerations. For example, mmWave on 60 GHz is prohibited in the US for use in the air. In consideration of drone development, the US Federal Communications Commission (FCC) has been carefully monitoring the activities. In August 2016, the FCC TAC (Technological Advisory Council) organized a meeting to collect input on UAS communication-related regulation. The "UAS" here is under broader definition and includes even low-orbit satellites. Many companies proposed a dedicated spectrum for control, while observations on using a cellular system were also shared. In 2018, the FCC initiated a TAC UAS Working Group which includes a Spectrum Subgroup. In the EU, similar activities are on-going. Recently [16], a new ECC report "Use of MFCN for UAS operations" was initiated, where 26 GHz is being considered for drone to ground communications.

Anticipating fast growth, R&D on drone technologies are encouraged by US federal agencies such as the FAA, Department of Transportation, and NASA. For example, the FAA allows special permits for testing new technologies, and several test sites have been established. Many companies have partnered with the FAA, such as CNN and PrecisionHawk, on different usages [17]. US NASA started an effort in late 2014 on UAS traffic management (UTM) [18]. The target is to enable large scale and safe drone deployments in low-orbital environments. In its vision, the system will coexist with today's manned aircraft management system, and will provide service at national scale. In November 2017, the US Department of Transportation announced a drone Integration Pilot Program (IPP) [19]. The purpose is to encourage local governments, industry, and academic partners to experiment with new innovations and usages on drones.

8.4.2 Academic Activities

There is also active research in academia on AVs, spanning from aerodynamic, communications, to economics. Here we focus on the wireless communication aspect.

A large portion of the research studied how to support drone communication/ control links using WiFi, 802.15.4, remote control channel (e.g. [20–24]) due to the spectrum availability. Aspects such as link quality, interference, requirements, antenna impacts, etc. have been considered. Many of the findings are relevant for cellular-based solutions as well. In particular, field experiments from [22, 23, 25] showed the importance of antenna design and pattern. It was shown that antenna nulls and direction could cause throughput loss. In [21, 22], the authors investigated WiFi protocols and discussed challenges due to mobility, heterogeneity, antenna, blockage, constrained resources, and limited battery power.

Reference [26] reviewed and compared the existing technologies, including 802.15.4, 802.11, 3G, LTE, and infrared. In [24], some of the drone applications (search and rescue, coverage, construction, and delivery) were compared in terms of constraints, assumptions, mission requirements, etc. Several works have also studied providing drone communication through wireless cellular infrastructure. Reference [27] investigated the interference problem when drones are present. It validated that the portion of interference coming from drone user equipment (UE) also depends on the 3D angle of arrival at the ground base station due to its antenna pattern and the interference to drones varies with altitude. In reference [28], cellular coverage was investigated in an isolated area. It was shown that the nulls in between side lobes create "holes in the sky."

A tutorial on drone-based wireless networks ([29] and references therein) has provided a comprehensive overview on the use of drones for wireless communications. The potential, state-of-the-art research achievement and challenges for a drone-based wireless system are discussed in the following areas: (i) coverage and capacity enhancement of beyond a 5G wireless network for supporting drone UEs, (ii) usage of drones as flying base stations for providing cellular service, (iii) integration on drones for Internet of Things communication, and (iv) usage of drones as flying backhaul to assist terrestrial networks.

8.4.3 3GPP Activities in General

Observing the rise in aerial markets, 3GPP initiated multiple activities to investigate how cellular networks should evolve to support reliable communication for diverse drone applications. Starting from defining drone use cases and service requirements, the 3GPP standard body identified the key challenges in supporting drones via LTE networks and further developed multiple candidate solutions to enhance LTE drone communications. Several working groups (WGs), mainly for the service and system aspects (SA) and radio access network (RAN) technical specifications, are involved in the cellular drone support discussion.

In the report of new service and market technology enablers for next generation mobile communication from 3GPP SA WG1 (services focus) [30], drones and robotics applications are identified as important usage cases in the category of critical communications. In [31], more in-depth descriptions on usages and requirements for drone applications are provided. The specification lists multiple drone usages with higher requirements on reliability, latency, service availability and/or positioning accuracy. In general discussion of service requirements for the 5G system [2], drones are also mentioned as new usages that can possibly impact 5G requirements.

In terms of supporting drone via LTE, the 3GPP RAN WG1 (physical-layer focus) and WG2 (medium access control and radio resource management layers focus) started a SI on enhanced LTE support for AVs in March 2017 [14]. The main outcome of the SI is captured in [3]. The report identifies key challenges in LTE drone support as well as providing an extensive list of candidate solutions for interference mitigation and mobility management for drones. Multiple companies provided simulation results and field trial data on LTE performance while supporting drones. After the SI completed in December 2017, a follow-up 3GPP RAN WI [15] continued to further enhance the LTE support for AVs. The WI successfully made several critical spec changes to enhance drone support, such as extending the parameter range for uplink power control, including 3D position information in the measurement report, height-based triggering for measurement reports, and signaling of flight path information.

In addition to supporting drone applications, 3GPP also investigated non-terrestrial networks (NTNs) that use satellites or air-borne vehicles such as high-altitude drones to provide 5G service. Multiple satellite and aerial access network architectures are identified. The study also discussed frequency bands, radio characteristics, propagation delay and Doppler modeling for NTNs. Detailed results from the study can be found in the summary report [32].

In 2018, 3GPP SA1 also started an SI on drone identification for safe drone operations [33]. In this study, new architecture and service designs are being considered for the core network to enable safer and secure operations.

8.5   3GPP Enhancement on Supporting Drones

In this section, we discuss more details on the observations and technology development in 3GPP for supporting drone communication. Several recent papers have also summarized the effort (e.g. [34]).

8.5.1   3GPP Drone Study Item and Work Item in RAN1

In this subsection, we focus on the 3GPP enhancement specification for drones in both SI and WI from the RAN1 perspective.

The SI aims to investigate the feasibility of serving AVs using LTE network deployments with base station antennas targeting terrestrial coverage, supporting Release 14 functionality (including active antennas and full-dimension multiple-input multiple-output [FD-MIMO]). The objectives of the study were to verify the level of performance, identify supportable heights, speed, and densities of AVs, select air-to-ground channel models, and study performance enhancing solutions for interference mitigation, etc.

Three deployment scenarios were considered in the SI for the system performance evaluation:

1) Urban-macro with aerial vehicles (UMa-AVs) where base station antennas are mounted above rooftop height of neighboring buildings in an urban environment.
2) Urban-micro with aerial vehicles (UMi-AVs) where base station antennas are placed below rooftop.
3) Rural-macro with aerial vehicles (RMa-AVs) where base station antennas are mounted on top of towers in a rural environment with larger inter-site distance separation.

In all the scenarios, drones are modeled as outdoor UEs with height uniformly distributed between 1.5 and 300 m above ground level (AGL).

For channel modeling, terrestrial users follow [35], while for AVs, new channel models are specified to characterize the channels between aerial UEs and eNodeBs regarding line-of-sight probability, pathloss, shadow-fading, and fast-fading. Some of the key parameters are explained in Table 8.1. Other details can be found in [3].

Table 8.1 Channel model assumptions for aerial vehicles.

Parameter	Assumption
Line-of-sight probability	Height-dependent and line-of-sight probability increases as height increases
Pathloss and shadow-fading models	Follow tables B-2 and B-3 in [3], respectively
Fast-fading model	Three alternatives are agreed
	The first one is based on a clustered delay line model
	The second one has adopted aerial UE height dependent modeling methodology for angular spreads, delay spreads, and K-factor
	The third one reuses the fast-fading model in [35] with the K-factor modified to 15 dB

Based on the above-mentioned scenarios and channel models, system performance is evaluated and increased network interference are observed in both uplink and downlink in the presence of aerial UEs. Due to the favorable line-of-sight propagation environment, aerial UEs in the sky can cause more uplink interference, and also suffer strong downlink interference from neighboring base stations. These problems will become more severe as the density of the aerial UEs increases, leading to significant overall signal-to-interference-plus-noise ratio (SINR) and throughput degradation. Hence, several interference mitigation methods are proposed including both implementation-based solutions and those requiring specification enhancements [3].

For example, to address downlink interference, intra-site joint transmission coordinated multipoint (JT-CoMP) can be applied where multiple cells belonging to the same site are coordinated and data is jointly transmitted to the UEs. Besides, coverage expansion has been proposed to enhance synchronization and initial access for aerial UEs.

As for mitigating uplink interference, uplink power control schemes with specification enhancement can be adopted, such as optimizing UE-specific open-loop power control parameters, and setting wider step size for the transmit power control command in the closed-loop power control.

In addition, FD-MIMO can be used without specification enhancement to mitigate both uplink and downlink interference. Employing directional antennas at aerial UEs is another effective approach in reducing the increased interference in both uplink and downlink, as demonstrated in [3].

In the WI following the completion of SI, new features are specified to further improve the efficiency and robustness of terrestrial LTE network to deliver connectivity solutions for drones. One of the objectives for RAN1 in the WI is to specify uplink power control enhancements. In LTE, fractional open-loop power control is introduced, where the UE transmit power per resource block is determined by

$$P_{tx} = \min\{P_{\max}, P_0 + \alpha \times PL\}[\text{dBm}],\tag{8.1}$$

where $P_{\max}$ is the maximum UE power, P_0 is a target received power level, α is the fractional pathloss compensation factor, and PL is the pathloss component. The current LTE spec supports only cell-specific α, cell-specific $P_0 \in [126, \cdots, 24]$ and UE-specific offset $\in [-8, \cdots, 7]$.

To enable more flexible uplink power control to cope with dynamic interference in the presence of drones, RAN1 has focused on the following areas for uplink power enhancement: (i) introducing UE-specific fractional pathloss compensation factor, and (ii) extending the range of UE-specific P_0 parameter. At the end of the WI, the following new features are introduced to Release 15:

- The range of UE-specific P_0 parameter is extended to $[-16, \cdots, 15]$.
- The aerial UE can be RRC configured with a single UE-specific α parameter for PUSCH.
- The UE-specific α overrides the cell-specific α once it is configured.
- The support of UE-specific α and the extended range of UE-specific P_0 are optional with capability signaling.

8.5.2 3GPP Drone Study Item and Work Item in RAN2

During the course of the WI for enhanced LTE drone support, RAN2 made multiple specification modifications to improve aerial communication links. The most important changes are related to airborne status indication, mobility management, interference detection and message overhead reduction. RAN2 defined signaling for indicating airborne or not, which is useful for identifying aerial users and then applying interference mitigation and mobility management enhancements. From the new spec, reference altitude information can be provided by eNB to aerial UE to assist UE on identifying its status.

From the RAN2 perspective, mobility management is the most critical challenge for reliable drone support. Simulation results [3] show aerial users 50–300 m above the ground experience higher handover failure rate than terrestrial users. As drone speed increases, all companies observe more frequent handover failure. Most results indicate that mobility management for drones needs to be improved in order to reliably support drone applications.

Fortunately, unique drone communication properties can be utilized to assist mobility management. Figure 8.2 shows how the reference signals received power (RSRP) from the upper-right sector of the central base station decays over an area covered by 57 cell sectors. The results are obtained based on 3GPP simulation assumptions [3] for different UE altitudes. The geographical heat map of UE wideband SINR over three cell sectors is plotted in Figure 8.3. Note that the shaded scales are all different. These figures show that channel quality for aerial link can fluctuate rapidly as drones move. There are regions in the sky with very poor channel quality to support satisfactory connection for drones. On the other hand, the statistics also suggest that aerial channels are less likely to be obstructed by objects. Compared with shadowing effect, the antenna pattern at the base station has more impact on aerial channel fluctuation. In other words, wireless link quality for drones is likely to be predictable based on location information. In addition, the height information can be useful to the network for selecting RAN parameters for drones. As a result, RAN2 approved to include 3D position information in the measurement report and enable drones to provide flight path trajectory to the network. New height-based measurement report triggering events, where vertical speed can be sent along with location information in the height report, are also introduced in Release 15 specifications.

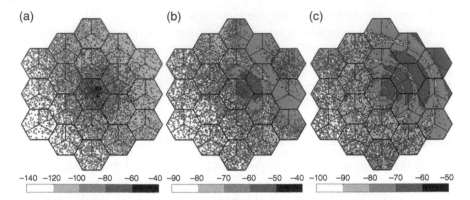

Figure 8.2 Geographical RSRP heat map at different altitudes: (a) ground level; (b) 100 m above ground; and (c) 300 m above ground. The plots show the RSRP from the upper-right sector of the central base station, as indicated by small triangles, measured over a geographical area of 19 cell sites (total 57 cell sectors).

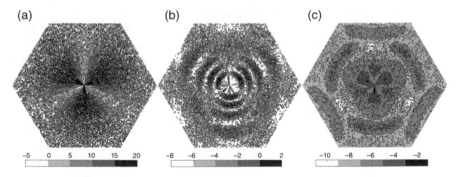

Figure 8.3 Geographical wide-band SINR heat map at different altitudes: (a) ground level; (b) 100 m above ground; and (c) 300 m above ground, over a geographical area of 1 cell site (3 cell sectors).

Another focus in RAN2 is interference detection mechanism and message over-head reduction for aerial UEs. Statistics from RAN1 study show that interference is the critical limiting factor to drone throughput. With enhanced interference detection mechanism, cellular systems can timely apply the right interference mitigation schemes developed from RAN1 study to improve drone connection. For example, a good interference detection mechanism can help quickly identify strong interfering neighbors to be muted when enhanced inter-cell interference coordination (eICIC) is used for aerial UEs. However, aggressive interference detection mechanism can cause excessive measurement report which is not desirable. In addition, as mentioned in Section 8.5.1, aerial users tend to observe more interfering neighbors due to the highly probable line-of-sight propagation environment which leads to unnecessary measurement reports triggering.

RAN2 thus targeted to adjust existing protocols to avoid excessive signaling overhead while maintaining best interference detection efficiency. Based on observations from aerial UE interference characteristics, RAN2 agrees to introduce a new parameter for measurement report triggering: the number of cells satisfying the condition of measurement trigger event. Only when an aerial UE observes at least the specified number of neighbor cells satisfying the triggering condition, a measurement report will be sent. The new multi-cell trigger mechanism can significantly reduce the number of measurement reports from aerial users.

8.6 5G Challenges, Solutions, and Further Studies

As 5G is emerging with advanced capabilities and more spectrum, it becomes a natural choice to cater for the demanding needs of aerial verticals. In this section, we expand the discussions on the unique properties of drones and elaborate on emerging new usages in the 5G era. Then we discuss how these properties link to key performance indicators (KPIs) such as throughput, reliability, positioning etc. and how existing 5G features help resolve the challenges. As AVs constantly evolve in terms of new usages, regulations, etc., we also discuss how future 5G could be continuously enhanced.

8.6.1 Challenges, New Emerging Usages, and Requirements for 5G

There are several unique and large verticals that 5G is trying to support, such as connected automobiles, connected industry, and connected health. AVs represents a new and emerging vertical with unique usages and requirements.

In Section 8.2 we identified several unique properties and challenges inherent to AVs: severe interference, irregular side lobes, mobility challenge, positioning need, and more focus on uplink. In the 5G era, these requirements continue to hold and will get tighter as the deployment scale and density continue to increase.

In addition to the above, at the beginning of 5G development, several new usages were considered in defining general 5G requirements. In 3GPP TR 22.862 "Enablers for Critical Communications," several use cases are discussed based on expected drone usages. These requirements were combined with other verticals and uses, and were characterized in the 5G requirement document TS 22.261 [2].

Particularly in TR 22.862, requirements focusing on latency, reliability, and availability were discussed based on several sets of usages. A major example of future drone uses is when a local team of drones collaborate as a sensor/actuator network. In this example, the drones need to work in uncertain and dynamic environments while being controlled by an operator. This could be used for, e.g. surveillance, collaborative mapping, and object manipulation. To enable such operations, one needs node-to-node communications, mesh networking, and end-to-end connectivity and reliability. This leads to the needs of continuous wireless coverage for UAV flying between 10 and 1000 m with maximum speed of 200 km/h. A 5 ms latency is needed for node-to-node links.

A second set of usages discussed in TR 22.862 centers around positioning requirements for low altitude UAV in critical condition. This includes package delivery, emergency surveying, and sensing by a team of drones. To satisfy these needs, high positioning accuracy, e.g. 0.5 m in both indoor and outdoor, along a density up to 1 million devices per square kilometer and with high mobility are needed [31].

Besides the above, new technology and uses of AVs are being developed constantly. There are several use cases in their early infancy that are already getting a lot of attention:

- Drone swarm on a large scale and with high density is one example. One use case is to deploy hundreds of drones for rescue after a natural disaster. Compared with a small team of drones as mentioned above in TR 22.862, large drone swarm use case expects drones to work more intelligently and collaboratively, in real time both locally and globally. This exerts stringent requirements on all aspects of wireless communications such as peer-to-peer, backhaul, and positioning and controls.
- Augmented reality and virtual reality applications with drones are another unique set of applications. In this use case, a team of drones can record videos of a scene in real-time, process and transmit data back to one or a set of end users for entertainment (e.g. Olympics) or for remote control of a rescue task in a dangerous environment. These tasks require optimized tradeoff between local processing on the device, edge computing for more complex tasks but short latency, and global optimization in a data center with longer delay.
- Another emerging usage of AVs is human-lifting "air-taxi" UAVs. Uber, Airbus, and Ehang are experimenting with UAVs for urban mobility, where a human can hire a drone as a taxi. Besides, there are also many startups experimenting with flying cars. This new usage sets extra requirements on reliability, safety, and computing support.

Besides these new use cases, concerns on public safety also introduce new requirements. Methods to identify a drone, to enforce air-space protection (e.g. geo-fencing) are urgently needed. Recently 5G SA1 approved a new item on studying drone IDs [33]. Several studies have come to realize that a cellular system, especially 5G, can serve as a "natural" infrastructure for providing such support.

Many of the aforementioned requirements can be addressed by focusing on enhanced basic KPIs such as throughput, latency, reliability, and positioning accuracy. Yet the connection between certain needs and basic KPIs may not be straightforward as new use cases develop.

8.6.2 3GPP Features Addressing These New Requirements

5G introduces many new features compared with 4G. These include particularly, among others, enhanced mobile broadband (eMBB), URLLC, and massive machine type communications (massive MTC). It also introduces a vast amount of spectrum in the mmWave band and enhanced V2X (vehicle-to-everything) support. At the core network side, network slicing via advanced network function virtualization (NFV) allows flexible and dedicated solutions toward supporting drones. Computing power for each local area is also enhanced for more powerful edge computing. These provide great potential in supporting AVs by addressing the challenges identified earlier.

In characterizing how 5G features can enhance drone support, one should keep in mind two very different types of links for AVs: command and control (C2), and data (Data). C2 is essential for drone operations such as fly control, collision avoidance, etc. It typically does not require high data rate, and the focus is reliability and low latency. Depending on the usage scenarios, requirements might be different. In 3GPP study, one typical assumption is that the data rate is 50–200 Kbps. On the other hand, the "Data" link is mainly needed for non-control traffic, e.g. video streaming. It needs large throughput but, depending on exact usage, may relax relatively on latency. In certain cases, such as augmented reality and virtual reality, both high data rate and low latency may be required.

8.6.2.1 eMBB

Technologies behind eMBB significantly enhance both C2 and Data links for AVs. First of all, eMBB provides links with extremely high data rate, e.g. Gbps, via technologies such as massive MIMO and mmWave. For AVs, this provides fast uplink and downlink for data intensive applications, helping answer the challenge to deliver air-to-ground high quality video streams. Flexible beamforming for both low band and mmWave allows good support even when drones are moving fast.

In both low band and mmWave bands, new technologies provide much sharper beams compared with 4G. With beam scanning and tracking interfaces, this alleviates the interference issues in 4G and earlier generations due to the broadcast nature of wireless communications. In particular, in low band, massive MIMO, a continuation of FD-MIMO, allows better resolution in the vertical domain, suited nicely for drones flying in 3D space. In mmWave band, besides much narrower beams, the form factor shrinks down significantly, allowing UE (such as drones) to be equipped with mmWave arrays for uplink beamforming and tracking.

Technologies behind eMBB also answer the C2 and positioning challenges of drones. Narrower beams with 3D beam scanning and tracking leads to less interruption for the control and command links. Wide bandwidth and dominant line-of-sight propagation inherent in mmWave allow more accurate positioning.

8.6.2.2 URLLC

5G provides new flexible interfaces to shorten the latency and improve reliability compared with 4G. In addition to HARQ, 5G introduces short transmission time interval (TTI), reducing from 1 to 0.125 ms in 5G NR [36]. For dynamic uplink scheduling or contention-based access, 5G now allows grant-free access without resource reservation via the network. The shortened response time provides solutions to mission-critical drone controls such as remotely controlled drone team/ swarm activities.

Reliability is enhanced several ways in 5G. Control channel design is very different compared with 4G. Now UE specific signaling based on downlink channel quality allows adaptation of MCS (modulation and coding set) to each UE's channel condition and latency budget. Multi-RAT between mmWave, low band, LTE, and other technologies allows broader diversity. For example, low band can be dynamically switched on when mmWave is blocked by an obstacle. These designs address the reliability requirement specified by 5G requirements in TR 22.862 for Critical Communications. For aerial vertical, it provides safety and reliability to support fast mobility in the air. This is more pronounced in the case when a drone is used as an air taxi.

8.6.2.3 Massive Machine-Type Communications

This provides services for a very large number of devices with relatively low requirement on, e.g. latency, and with high density (e.g. 100 million devices per square kilometer). These devices typically need to have long battery life and with low cost – supporting the vision of the Internet of Things. To address these special needs, 5G designs provide several solutions. For example, 5G allows grant-free and asynchronous access, shorter TTI, and massive MIMO. These considerations and designs conveniently help the aerial vertical usage as well. They are suited for geofencing, large drone deployment with low data rate (e.g. Control and Command), drone swarms, drone IDs, and drone operations over large sensor networks.

8.6.2.4 V2X

Continuing existing work from LTE, 5G has been enhancing V2X designs in sync with the revolution of autonomous driving. Release 14 provides enhanced range and reliability with high density support. It also reuses DSRC/C-ITS higher layers, and, for lower layer, V2X can coexist with 802.11p on the ITS band. As an integral component, V2X provides support on safe operations such as collision warning, sensor sharing, intention/trajectory sharing, wideband raging and positioning, and local maps. Although these new features were targeted for cars, they naturally enhance drone support in the 5G era. Nevertheless, further evaluations and new design considerations need to be carefully done for drones under the corresponding channel models and deployment.

8.6.2.5 Next Gen Core Network

5G core network designs provide true flexibility in supporting many use categories. This is achieved by software defined networking (SDN), NFV, network slicing, and cloud RAN. Network slicing allows different network segments to be configured as needed for providing differentiated end-to-end services. This provides a framework for optimized AV support in cross-layer and end-to-end fashion. Furthermore, a private network may be an option for dedicated service.

8.6.2.6 Positioning

Positioning is essential for AV operations in 3D space, especially for safety. For example, in geo-fencing, drones need to operate within a specified region or not to enter a protected region. In multi-drone operation and autonomous operation, positioning is a basic function for collision avoidance or cooperation, between drone to drone or drone to other objects. 5G continues to enhance higher resolution of UE positioning. Sub-meter accuracy is expected with wider bandwidth and better signaling. This, combined with advanced features from V2X such as interfaces for vehicles to share their positions more accurately and frequently, will enhance AV operations dramatically.

8.6.3 Further Study Needed for Aerial Vehicles in 5G

Even though current 5G features seem capable of solving many challenges for supporting AVs, unique properties of drone applications were not carefully considered when basic 5G requirements and structures were established. Most 5G designs have been focusing on ground users in contrast to aerial users in 3D space with high mobility. Therefore, new studies are required in future 5G evolution for better support of aerials.

As discussed earlier, 5G RAN1 and RAN2 support both FD-MIMO and massive MIMO. Yet, the parameters for the control and basic public channels were designed and evaluated mostly based on UEs on or close to the ground. Given excessive interference from drones in the air, the capacity of such channels may be limited. Careful study on such basic channels with possible enhancement in the vertical domain is needed. The tradeoff between supporting aerial users and terrestrial users is a key concern. More flexible design allowing real time adaptation may be needed depending on the combined traffic pattern.

Another challenge for eMBB regarding drones is the applicability of mmWave on airborne drones. Current regulations are very restrictive on mmWave transmission for a mid-air vehicle. Future regulations need to be studied and relaxed for aerial usage. Currently US FCC and EU agencies have started investigating such possibility. For example, 26 GHz has been discussed for possible drone to ground usage in the EU.

Release 15 drone study introduced RAN2 enhancement for supporting drones. For example, drones can provide flight information to the network, so that the latter may utilize it to optimize communication support accordingly. Even though discussions were held for height-based optimization, coherent conclusions on how to enhance support were not reached. Future research remains to be done to best utilize these new features in Release 15 and to discover new mobility enhancement strategies.

Regarding URLLC to AVs, more careful designs should be considered. The aviation community needs to be brought in for addressing aviation considerations such as traffic management and coordination with manned aircraft. For example, in normal aerial operation, safety mechanisms are not designed based on assuming a ms-level communication latency. Many on-board mechanisms, such as collision avoidance via sensing, are in place for safe drone operation. On the other hand, for future usages such as high-density drone operation with remote control, such low latency is required. URLLC with specialization for UAVs is needed to support different use scenarios such as: (i) safe drone operation with single unit, (ii) drone operation with multiple or dense drones, and (iii) drones as tools for other usages such as in rescue tasks.

In practice, several generations of cellular network coexist at the same time, and there are coverage issues in rural areas. Wireless support for drones needs to carefully consider coverage issues and multi-RAT. Combinations between 5G with 4G or even 3G will provide different levels of support. Multi-RAT with, e.g. PAN/LAN, or their specialized versions, is needed for safe aerial operations with reliability and redundancy. Furthermore, the UTM system has been considered by many to be the future drone operation model, in coexistence with manned air traffic. Combining UTM and 5G may provide benefit for both sides. Actually, 5G has been considering drone ID in SA1 study [33] and integration of aerial support functions to core network can be one option. A good question is how one should split the solutions between radio link, core network, and over-the-top application layers.

As safety is a key concern for aerial operations, 5G and earlier generations can be utilized to provide further support for safe operation. For example, mmWave equipment on the base station towers entails a possibility of new signaling and feedback design for detecting objects in the air similar to radars, thus providing a new functionality besides communication.

Finally, AVs can also be part of 5G infrastructure. As they can be deployed on demand and to areas without fixed infrastructure, drones can be used to help improve/recover cellular coverage. They can serve as flying cells, backhaul, or relays in situations such as disaster recovery and sport events. With a non-conventional 3D network in mind, new designs and investigations are definitely needed.

Acronyms

3GPP	Third Generation Partnership Project
CTIA	Cellular and Telecommunications and Internet Association
ECC	European Electronic Communications Committee
FAA	Federal Aviation Agency
FCC	Federal Communications Commission
GSMA	GSM Association
LTE	Long Term Evolution
MCS	Modulation and coding set
NASA	National Acronautics and Space Administration
PAN	Personal area network
PUSCH	Physical uplink shared channel
QoS	Quality of service
RAN	Radio access network
RC	Remote control
SI	Study Item
TTI	Transmission time interval
UAV	Unmanned aerial vehicle
UE	User equipment
WAN	Wide area network
WI	Work Item

References

1 Uber Elevate (2016). The future of urban mobility. https://www.uber.com/info/elevate (accessed 14 October 2019).
2 3GPP TS22.261 (2018). Service requirements for next generation new services and markets.
3 3GPP TR36.777 (2018). Enhanced LTE support for aerial vehicles.
4 FAA (2018). Part 107. https://www.faa.gov/uas/media/Part_107_Summary.pdf (accessed 21 September 2018).
5 China Civil Aviation Bureau (2015). Interim regulations on UAS.

6 Qualcomm (2016). Qualcomm and AT&T trial drones on cellular network. https://www.qualcomm.com (accessed 14 October 2019).

7 Qualcomm (2017). LTE unmanned aircraft systems, trial report v1.0.1, May 2017.

8 Qualcomm (2016). Accelerating integration of drone into the cellular network. FCC TAC Meeting, August 2016.

9 Qualcomm (2018). Leading the world to 5G: Evolving cellular technologies for safer drone operation. https://www.qualcomm.com/documents/leading-world-5g-evolving-cellular-technologies-safer-drone-operation (accessed 26 September 2018).

10 Donovan, J. (2018). Drones taking our network to new heights. http://about.att.com/innovationblog/drones_new_heights (accessed 26 September 2018).

11 The GSMA (2018). Mobile-enabled unmanned aircraft: How mobile networks can support unmanned aircraft operations.

12 Cellular Telecommunications and Internet Association (2017). Commercial wireless networks: the essential foundation of the drone industry.

13 Civil Aviation Administration of China (2018). Low-altitude connected drone flight safety test report, January 2018.

14 NTT, Docomo, Ericsson (2017). 3GPP RP-170779 Study on Enhanced Support for Aerial Vehicles, March 2017.

15 Ericsson (2017). RP-172826 New WID on Enhanced LTE Support for Aerial Vehicles, December 2017.

16 ECC (2018). PT1 Meeting #59, Edinburgh.

17 FAA (2019). Programs, Partnerships & Opportunities. https://www.faa.gov/uas/programs_partnerships (accessed 14 October 2019).

18 NASA (2015). NASA UTM. https://utm.arc.nasa.gov (accessed 21 September 2018).

19 US FAA and Department of Transportation (2018). UAS Integration Pilot Program. https://www.faa.gov/uas/programs_partnerships/uas_integration_pilot_program (accessed 28 September 2018.

20 Gupta, L., Jain, R., and Vaszkun, G. (2016). Survey of important issues in UAV communication networks. *IEEE Communication Surveys and Tutorials* 18: 1123–1152.

21 Van den Bergh, B., Vermeulen, T., and Pollin, S. (2015). Analysis of harmful interference to and from aerial IEEE 802.11 systems. *Proceedings of the First Workshop on Micro Aerial Vehicle Networks, Systems, and Applications for Civilian Use, DroNet'15*, New York, NY, USA, pp. 15–19. ACM.

22 Asadpour, M., Van den Bergh, B., Giustiniano, D. et al. (2014). Micro aerial vehicle networks: an experimental analysis of challenges and opportunities. *IEEE Communications Magazine* 52: 141–149.

23 Cheng, C., Hsiao, P., Kung, H.T. et al. (2006). Performance measurement of 802.11a wireless links from UAV to ground nodes with various antenna orientations. *Proceedings of the 15th International Conference on Computer Communications and Networks*, pp. 303–308.

24 Hayat, S., Yanmaz, E., and Muzaffar, R. (2016). Survey on unmanned aerial vehicle networks for civil applications: a communications viewpointe. *IEEE Communication Surveys and Tutorials* 18: 2624–2661.

25 Ahmed, N., Kanhere, S.S., and Jha, S. (2016). On the importance of link characterization for aerial wireless sensor networks. *IEEE Communications Magazine* 54: 52–57.

26 Andre, T., Hummel, K.A., Schoellig, A.P. et al. (2014). Application-driven design of aerial communication networks. *IEEE Communications Magazine* 52: 129–137.

27 Bergh, B.V.D., Chiumento, A., and Pollin, S. (2016). LTE in the sky: trading off propagation benefits with interference costs for aerial nodes. *IEEE Communications Magazine* 54: 44–50.

28 Teng, E., Falcao, D., Dominguez, C. et al. (2015). Aerial sensing and characterization of three-dimensional RF fields. Second International Workshop on Robotic Sensor Networks, Seattle, USA.

29 Mozaffari, M., Saad, W., Bennis, M. et al. (2018). A tutorial on UAVs for wireless networks: Applications, challenges, and open problems, CoRR, vol. abs/1803.00680.

30 3GPP TR22.891 (2016). Study on new services and markets technology enablers.

31 3GPP TR22.862 (2016). Feasibility study on new services and markets technology enablers for critical communications.

32 3GPP TR38.811 (2018). Study on New Radio (NR) to support non-terrestrial networks.

33 3GPP TR22.825 (2018). Study on remote identification of Unmanned Aerial Systems (UAS).

34 Lin, X., Yajnanarayana, V., Muruganathan, S.M. et al. (2018). The sky is not the limit: LTE for unmanned aerial vehicles. *IEEE Communications Magazine* 56: 204–210.

35 3GPP TR38.901 (2018). Study on channel model for frequencies from 0.5 to 100 GHz.

36 Shafi, M., Molisch, A.F., Smith, P.J. et al. (2017). 5g: a tutorial overview of standards, trials, challenges, deployment, and practice. *IEEE Journal on Selected Areas in Communications* 35: 1201–1221.

9

Connected Automobiles

Murali Narasimha[1],* and Ana Lucia Pinheiro[2],**

[1] *Futurewei Technologies, Rolling Meadows, IL, USA*
[2] *Intel Corporation, Hillsboro, OR, USA*
* *Current affiliation: Murali Narasimha, Intel Corporation, Hillsboro, OR, USA*
** *Current affiliation: Ana Lucia Pinheiro, Samsung Electronics America, Plano, TX, USA*

Abstract

This chapter focuses on the value that 5G brings to connected vehicles. In particular, the higher data rates and lower latency enable in-vehicle applications that can consume large amounts of data, while low-latency communication can enable fast mechanical control loops. In addition to high data rates and low latency, one key feature of 5G that is particularly relevant to connected vehicles is edge computing. By deploying computing capabilities near the end user, the system can take advantage of localized information, such as map updates and weather notifications. This chapter concentrates in two main use cases: vehicle platooning and high definition maps, and how 5G and edge computing can assist to provide the best experience to the end user. Before discussing the use cases, the chapter describes the five levels of vehicle automation and give an overview of multi-access edge computing, an important feature of 5G that helps enable use cases such as the ones described.

Keywords *5G network; connected automobiles; high definition maps; multi-access edge computing; platoon-based driving; vehicle automation*

9.1 Introduction

Anyone who has lived in or near a metropolitan area is aware that our current transportation systems are stretched beyond capacity. The loss of life and damage to property from automobiles is significant [1] and the resulting economic impact is

5G Verticals: Customizing Applications, Technologies and Deployment Techniques, First Edition.
Edited by Rath Vannithamby and Anthony C.K. Soong.
© 2020 John Wiley & Sons Ltd. Published 2020 by John Wiley & Sons Ltd.

staggering [2]. Additionally, lost productivity due to traffic congestion shows the economic toll on large population centers [3–5]. The US National Highway Traffic Safety Administration (NHTSA) has observed that 94% of vehicular accidents are caused by human error [6]. Of these, 41% are attributed to recognition errors (resulting from inattention, inadequate surveillance, distractions, etc.) and 33% are attributed to decision errors (incorrect assumptions about others' actions, speeding, etc.).

Vehicular transportation has not seen major innovations since the introduction of the modern gasoline engine based vehicles. It is widely recognized now that some of the breakthroughs that enabled pervasive connectivity (such as high bandwidth wireless communication, positioning, machine learning), combined with other emerging technologies can be used to gradually decrease the human involvement in driving. Automation of driving has become perhaps the most significant technology goal of the first half of the twenty-first century. Two technology areas that form the underpinnings of automated vehicles are:

- Sensing: Sensors such as radar and lidar, and also image recognition systems are necessary for a vehicle to learn of its surroundings.
- Robotics: Planning and executing sequences of precise mechanical actions is necessary for automated driving.

Automated vehicle technology has moved from being a research activity in the area of Robotics to mainstream. Leading technology companies such as Google, Tesla, and Uber are engaged in developing various types of automated vehicle solutions. Most vehicle manufacturers have their own plans and roadmaps toward automated driving and are engaged in various technology trials. Some Advanced Driving Assistance Systems (ADAS) are now available in vehicles (e.g. adaptive cruise control (ACC), lane tracking, emergency braking).

Much of the current generation of automated vehicle technology relies on sensors and associated processing to learn about the changing environment. Cameras and image processing are used to identify traffic situations (e.g. traffic signals), pedestrians, animals, etc. Global Positioning System (GPS) provides absolute positioning information. The data provided by the sensors are inputs to a complex system that determines the appropriate steps to take (course correction, braking, acceleration, etc.) and activates the relevant mechanical controls.

Although sensors are essential for automated driving features, they are hardly a complete solution. The following characteristics generally apply to all sensors:

- *Sensors have a short range.* Radar, lidar, and cameras all have a line-of-sight coverage. As a result, environmental features such as buildings, trees, large vehicles, etc. pose a significant problem since the sensors cannot see past obstructions.
- *Sensors can fail to detect problematic situations.* Different sensors have different specific weaknesses. Lighting conditions can affect the performance of cameras and weather conditions can affect lidar and radar. Most vehicles that advertise

ADAS and self-driving features today require the driver to be always attentive and able to take control of the vehicle in case of an unexpected event.

- *Sensors neither improve traffic flow nor reduce congestion.* A vehicle equipped with sensors can be automated (to a degree dependent on the types of sensors used) but still has no information about the intentions of other vehicles in its proximity. Consequently, even if a vehicle is equipped with the most precise sensors, it essentially has to emulate a human driving the vehicle and be prepared for any maneuver that could be performed by a vehicle or pedestrian in the vicinity. This may give the vehicle occupant some comfort by reducing their involvement in driving but it does not solve the larger issues of optimizing traffic flow, reducing travel times, and reducing transportation related energy consumption.

These limitations of sensors have led to studies that envision a combination of sensing and wide-area communication to enable higher degrees of automation [7, 8]. The exchange of messages between vehicles and infrastructure or direct communication among vehicles may be very beneficial to enhance an ADAS decision making process. The concept of ADAS is illustrated in Figure 9.1. The messages can deliver sensor information recorded by vehicles, and also information locally stored in the vehicle such as high definition (HD) maps.

This chapter focuses on the value that 5G brings to connected vehicles. In particular, the higher data rates and lower latency enable in-vehicle applications that can consume large amounts of data, while low-latency communication can enable fast mechanical control loops. In addition to high data rates and low latency, one key feature of 5G that is particularly relevant to connected vehicles is edge computing. Edge computing enables placing intelligence at locations in the network close to the user. The low latency combined with the edge computing

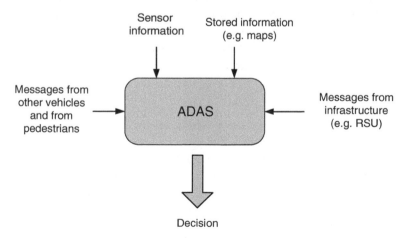

Figure 9.1 Information that may be utilized by an Advanced Driving Assistance System.

capabilities can enable intelligent and fast mechanical control. The high data rates combined with edge computing can make ADAS features more intelligent. By deploying computing capabilities near the end user, the system can take advantage of localized information, such as map updates and weather notifications.

This chapter concentrates on two main use cases: vehicle platooning and HD maps, and how 5G and edge computing can assist to provide the best experience to the end user. Before discussing the use cases, we describe the five levels of vehicle automation and give an overview of multi-access edge computing (MEC), an important feature of 5G that helps enable use cases such as the ones described.

9.2 Levels of Vehicle Automation

In light of the rapid progression toward driving automation, governments and regulatory bodies are grappling with how transportation systems of the future will function. NHTSA has defined five levels of vehicle automation [9] which can be summarized as follows:

- No Automation (Level 0).
- Function-specific Automation (Level 1): one or more specific control functions (e.g. electronic stability control).
- Combined Function Automation (Level 2): automation of at least two control functions designed to work in unison (e.g. ACC in combination with lane centering).
- Limited Self-Driving Automation (Level 3): enable driver to cede full control of all safety-critical functions under certain traffic or environmental conditions; driver is expected to be available for occasional control (e.g. Google car).
- Full Self-Driving Automation (Level 4): vehicle performs all safety critical driving functions and monitors roadway conditions for entire trip; driver may not be available for control at any time during the trip and includes unoccupied vehicles.

While each successive level of automation represents significant technological advancement, the shift from Level 3 to Level 4 is especially challenging. While Level 3 automation relies on sophisticated interaction between various systems within the vehicle, the input for decision making is the environmental information obtained from the sensors. Level 4 brings in more challenging scenarios which may not be foreseeable by the sensors. In particular Level 4 automation requires: (i) environmental knowledge beyond the limits of the sensors on the vehicle, and (ii) combining and interpreting different streams of information to make and implement decisions as well as or better than a human driver. The usage of up-to-date HD maps is fundamental for Level 4. HD maps will be further discussed in a later section in this chapter.

9.3   Multi-Access Edge Computing in 5G

An MEC system is characterized by deploying compute and storage resources closer to end-point devices. Benefits of MEC include reduction of application latency, improvement of service capabilities, facilitation of data privacy requirements, and optimization of network utilization and cost. By placing content near the end user, the content can be accessed much faster, and without increasing the bandwidth requirements toward the backend cloud. In addition, applications can have access not only to local content but also to real-time information related to the network conditions. Typical information that can be exchanged between the 5G network and the MEC platform is information related to user equipment (UE) events, such as UE loss of connectivity, UE reachability, UE location, and even predicted UE movement and communication characteristics [10]. MEC requirements and capabilities are currently being standardized by the European Telecommunications Standards Institute (ETSI) [11].

There are a few different ways to deploy MEC in a cellular network. In one approach, a typical edge server may be deployed right behind the 5G base station, allowing for a break-out of data without the need to go through the core network. In another approach, the edge server may be deployed behind the 5G core network. Both approaches are valid; it is up to the network operator to decide on the best deployment choice given their current network implementation. Once deployed, each edge cloud platform may provide services to users in a given geographical area. The geographical area serviced by any given edge cloud platform may include one or more cells (i.e. one or more base stations). Figure 9.2 depicts a typical deployment of an MEC system.

As a consequence of hosting the application near the edge, the system needs to be capable of handling user mobility. Since edge servers may be attached to many different base stations, the system needs to decide if and when to migrate services to another edge server as the user moves. Once the decision is made, the type of application will drive the requirements for the migration:

- Stateless applications: A stateless application is an application that does not memorize the service state or recorded data about the user for use in the next service session.
- Stateful applications: A stateful application is an application that can record the information about the service state during a session change.

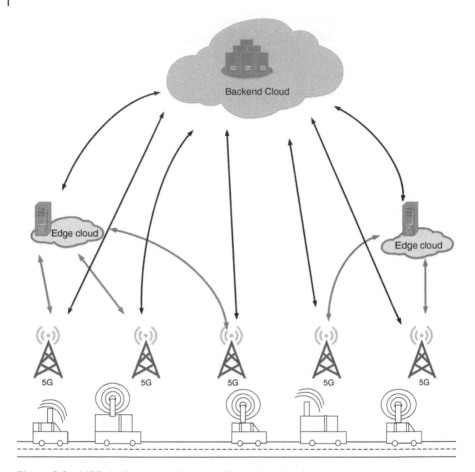

Figure 9.2 MEC platforms servicing specific geographical areas.

Stateful applications require transferring the user's context and/or application instance from one MEC host to another in order to continue offering an optimized service experience for the user. Stateless applications, on the other hand, do not require service migration.

The MEC application mobility feature is a work in progress in ETSI MEC ([12, 13]). Next, we discuss how 5G can support platoon-based driving.

9.4 Platoon-Based Driving Use Case

Platooning is one of the most investigated applications for connected vehicles. Many vehicles today come equipped with ACC which enables a driver to follow another vehicle with a desired following distance. This is done with the use of

Figure 9.3 Demonstration of an eight sedan platoon. *Source:* Photo courtesy of California PATH.

radar or ultrasonic sensors to measure the distance to the vehicle being followed and accelerating or decelerating as needed to maintain the desired distance. The goal of platooning is to extend the use of ACC such that a string of vehicles moves as a platoon.

A platoon of vehicles is defined as a train of vehicles moving along a roadway with one lead vehicle, as shown in Figure 9.3. Each vehicle follows the vehicle in front of it maintaining a safe distance. The lead vehicle may have a human driver fully in control and drivers of the vehicles following may be less actively involved in driving. Thus even with only limited ADAS capabilities (sensors to measure distances to preceding vehicles), significant self-driving capabilities can be achieved in scenarios with highway travel with straight roadways. The addition of lateral control and lane tracking can enable approximately Level 3 self-driving on highways. Platoon driving also enables more fuel efficiency (due to reduced aerodynamic drag) and can also lead to more efficient traffic flow due to being able to pack vehicles closer together with less likelihood of collisions.

In order to support platoon operations, ACC is extended such that vehicles communicate with each other; in particular the lead vehicle in the platoon provides relevant information to the following vehicles. This is known as cooperative adaptive cruise control (CACC). In this section we briefly describe the history and the current state of the art in CACC-based vehicle platooning. We describe the key underlying principles and the role 5G communication can play in this technology.

Sensors such as radar and ACC are the key requirements for platooning. ACC performs electronic actuation of the engine and brakes, based on input from radar or other range estimation systems. Each vehicle in a platoon has a controller that takes as inputs information such as the distance to the preceding vehicle, own speed and acceleration, preceding vehicle's speed, acceleration and intended action, and lead vehicle's speed and acceleration.

Platooning has been the subject of study for several decades from a theoretical standpoint in the automotive community. More recently, there have been various trials that use the ideas that have been developed to demonstrate practical platooning approaches. The work in [14] describes an ACC system that uses a safety distance separation rule between vehicles that increases vehicle throughput without causing oscillations in inter-vehicle spacings. In [15], the authors report results of one of the first systematic studies of platoon behavior by using a nonlinear simulation model and described the design of a control system for platoons.

There have been many trial activities focused on platooning. The trucking industry in particular has substantial interest in platooning technologies as it can enable fuel savings due to reduced aerodynamic drag for the following vehicles (in addition to all the other advantages of platooning) [16]. Major truck manufacturers are engaged in demonstration of on-highway truck platooning (for example see [17]).

9.4.1 Platoon Model

A mathematical model is provided below for describing a CACC-based platoon, and is illustrated in Figure 9.4. $\mathcal{L}$ denotes the length of each vehicle, v_i denotes the velocity of the ith vehicle in the platoon and g_i denotes the spacing between the ith vehicle and the $(i-1)$th vehicle. A vehicle follows a preceding vehicle in the platoon at a constant time headway. The time headway represents the duration the vehicle would take to cover a distance at its present speed. It captures the intuitive notion that the gap should be larger at higher speeds to accommodate longer stopping distances at higher speeds. That is, the desired inter-vehicle

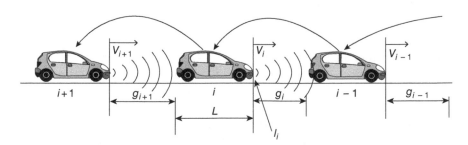

Figure 9.4 Platoon system parameters.

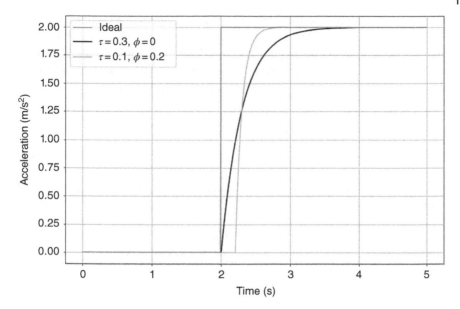

Figure 9.5 Vehicle response modeling.

gap for the ith vehicle $g_{des,i}$ is defined in terms of a constant time headway h and the current velocity:

$$g_{des,i}(t) = g_{min} + hv_i(t) \tag{9.1}$$

where g_{min} denotes a minimum gap. The *gap error* with respect to the preceding vehicle is $e_i(t) = g_i(t) - g_{des,i}(t)$.

Vehicles respond to acceleration and deceleration commands with a time lag. This is modeled as in the following. The external input to the engine (i.e. the desired acceleration) and the actual acceleration are related as follows:

$$\tau \dot{a}_i(t) = -a_i(t) + u_i(t) \tag{9.2}$$

where u_i is the desired acceleration of the ith vehicle, a_i is the acceleration of the ith vehicle, $\dot{a}_i$ is the derivative of the acceleration of the ith vehicle, and τ is a constant representing engine dynamics. Additionally, vehicles can have a time delay ϕ before responding to the input command. Thus the vehicle response can be represented as:

$$\tau \dot{a}_i(t) + a_i(t) = u_i(t - \phi) \tag{9.3}$$

where ϕ is a time delay before the vehicle response to the command begins.

Figure 9.5 illustrates the vehicle response to an application of a step input increasing acceleration by 2 m/s², for different values of τ and ϕ. Larger vehicles have a larger value of the time lag constant τ, which results in a longer time to achieve the desired acceleration.

The controller at each vehicle tries to drive the gap error to zero by applying acceleration or deceleration commands. We illustrate below mathematically how the gap error is minimized utilizing information from the preceding vehicle [15, 18]. Note that $v_i(t) = \dot{l}_i(t)$ and $a_i(t) = \ddot{l}_i(t)$, where $l_i(t)$ is the location of the ith vehicle at time t. We write the gap error and its derivatives as follows:

$$e_i(t) = (l_{i-1}(t) - l_i(t) - \mathcal{L}) - (g_0 + hv_i(t)) \tag{9.4}$$

$$\dot{e}_i(t) = \dot{l}_{i-1}(t) - \dot{l}_i(t) - ha_i(t) \tag{9.5}$$

$$\ddot{e}_i(t) = \ddot{l}_{i-1}(t) - \ddot{l}_i(t) - h\dot{a}_i(t) \tag{9.6}$$

From Eq. (9.3) we have $\dot{a}_i(t) = -\frac{1}{\tau} a_i(t) + \frac{1}{\tau} u_i(t - \phi)$. Substituting this into Eq. (9.6) yields:

$$\ddot{e}_i(t) = a_{i-1}(t) - a_i(t) + \frac{h}{\tau} (a_i(t) - u_i(t - \phi)) \tag{9.7}$$

Taking the next derivative of $e_i(t)$ and simplifying we have:

$$\dddot{e}_i(t) = -\frac{1}{\tau} \ddot{e}_i(t) - \frac{1}{\tau} C_i + \frac{1}{\tau} u_{i-1}(t - \phi) \tag{9.8}$$

where $C_i = hu_i(t - \phi) + u_i(t - \phi)$ represents the action taken by the ith vehicle. Note the dependence of $\dddot{e}_i(t)$ on u_{i-1} (the desired acceleration at the $(i-1)$th vehicle). This is due to dependence of the gap error derivatives on a_{i-1} (acceleration at the $(i-1)$th vehicle). C_i can be viewed as correcting for the observed gap errors and the observed result of the control input u_{i-1} to the preceding vehicle.

A standard PID controller is assumed and a control law is chosen as follows [19]:

$$C_i \triangleq k_1 e_i(t) + k_2 \dot{e}_i(t) + k_3 \ddot{e}_i(t) + u_{i-1}(t) \tag{9.9}$$

where k_1, k_2, and k_3 are constants. The action performed by the preceding vehicle u_{i-1} is received in a message from the preceding vehicle. It is assumed here that the delay in receiving the message from the preceding vehicle is negligible; in general a communication delay of θ can be reflected in Eq. (9.9) by replacing $u_{i-1}(t)$ with $u_{i-1}(t - \theta)$.

Using a vector $\mathbf{e}_i(t) = [e_i(t) \quad \dot{e}_i(t) \quad \ddot{e}_i(t)]^T$ to represent the gap error and its derivatives, the following closed loop system is obtained:

$$\dot{\mathbf{e}}_i(t) = \begin{pmatrix} 0 & 1 & 0 \\ 0 & 0 & 1 \\ -\dfrac{k_1}{\tau} & -\dfrac{k_2}{\tau} & -\dfrac{(1+k_3)}{\tau} \end{pmatrix} \mathbf{e}_i(t) + \begin{bmatrix} 0 \\ 0 \\ \dfrac{1}{\tau} \end{bmatrix} (u_{i-1}(t - \phi) - u_i(t)) \tag{9.10}$$

$$\dot{u}_i(t - \phi) = \begin{bmatrix} \dfrac{k_1}{h} & \dfrac{k_2}{h} & \dfrac{k_3}{h} \end{bmatrix} \mathbf{e}_i(t) - \frac{1}{h} (u_i(t - \phi) - u_{i-1}(t)) \tag{9.11}$$

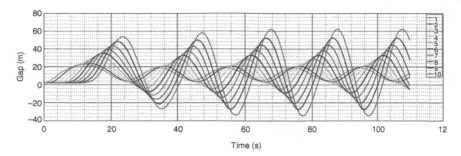

Figure 9.6 Inter-vehicle gaps with ACC.

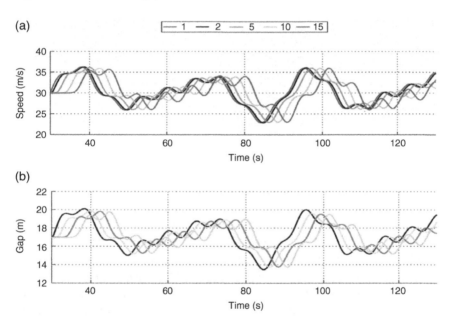

Figure 9.7 (a) Speed and (b) gap for vehicles 2, 5, 10, and 15 in a platoon at 30 m/s average speed and 13.33 Hz message rate (1 represents lead vehicle) [20].

The above closed loop system enables the ith vehicle to compute and apply the desired action to minimize the gap error; that is $u_i(t + \Delta t) = u_i(t) + \Delta t\, \dot{u}_i(t)$.

Figures 9.6 and 9.7 compare the simulated gap performance of ACC-based platoons with CACC-based platoons. The model for ACC is obtained by setting the u_{i-1} term in Eq. (9.9) to zero (representing the absence of information about the action taken by the preceding vehicle). The scenario consists of a lead vehicle with a sinusoidally time-varying speed with an average speed of 20 m/s, followed by nine vehicles in the platoon. A time headway $h = 0.5$ s, minimum gap $g_0 = 2$ m and $[k_1\, k_2\, k_3] = [0.2\, 0.7\, 0]$ are assumed. As can be seen from Figure 9.6, the inter-vehicle gaps increase rapidly toward the back of the platoon, clearly demonstrating an instability in the system.

(a)

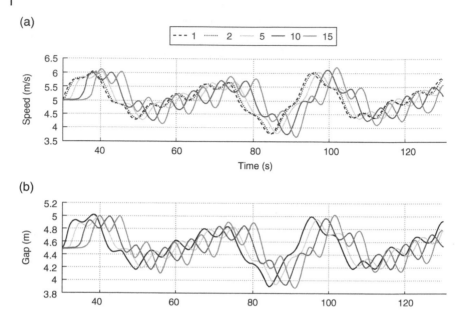

Figure 9.8 (a) Speed and (b) gap for vehicles 2, 5, 10, and 15 in a platoon at 5 m/s average speed and 5 Hz message rate (1 represents lead vehicle) [20].

Figures 9.7 and 9.8 ([20]) show the simulated performance of a CACC-based platoon. The scenario consists of long platoons of 15 vehicles on a 4-lane highway. A time headway of 0.5 s is assumed. Messages are transmitted through a base station and a communication delay of up to 100 ms is assumed (in line with Long-Term Evolution [LTE] network communication). All vehicles transmit 300 byte messages 13.33 times per second, with the platoon moving at an average speed of 30 m/s. The bandwidth supported by LTE limits the number of vehicles. At an average platoon speed of 30 m/s, approximately 312 vehicles are associated with a base station and a message rate of 13.33 Hz can be supported. Figure 9.7 shows the speeds and gaps of the vehicles for the 30 m/s case. As the speed decreases, the vehicle density increases. At an average platoon speed of 5 m/s, approximately 870 vehicles are associated with a base station and the message rate that can be supported falls to 5 Hz. Figure 9.8 shows the speeds and gaps of the vehicles for the 5 m/s case. Although in both cases the gaps are within safe limits, the second case shows larger variations of speeds and gaps of following vehicles compared with the lead vehicle.

The LTE bandwidth also limits the sizes of messages transmitted by the vehicles. The 300 byte size assumed above corresponds to a relatively simple Basic Safety Message. If additional information is transmitted (e.g. vehicle information,

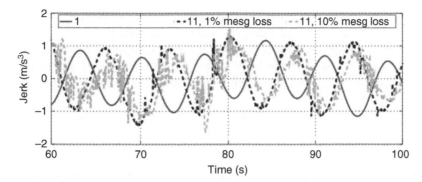

Figure 9.9 Comparison of jerk at the 11th vehicle in a platoon at 1 and 10% message error rates [20].

sensor data, etc.), the message sizes can be much larger. This then limits the message rates, which in turn affects the platoon performance.

Reliability of the communication has an impact on passenger comfort. One of the measures of passenger comfort is the derivative of the acceleration, also known as jerk. Figure 9.9 ([20]) shows that higher message error rates can significantly increase jerk and result in passenger discomfort. Furthermore, there is a need to have very short time headways for truck platooning, to minimize aerodynamic drag and maximize fuel efficiency. While LTE can support reliable communication mainly by using techniques such as more repetitions or higher transmit power, such techniques come at the expense of resource efficiency and as a result limit the number of vehicles that can be supported.

While achieving small or zero gap errors is the main underlying principle in CACC-based platooning, there are many other challenges to safe platoon operation:

- Lateral control: In addition to the longitudinal control described above, it is necessary to have lateral control such that vehicles stay in lane and follow curved roadways in a platoon.
- Vehicle dynamics and actuation delays: Mechanical responses have time delays and typically exhibit nonlinear behavior. Furthermore, mechanical characteristics of vehicles can change over time.
- Responding to abrupt braking and stops: Given different braking capabilities of vehicles, abrupt speed reductions of vehicles in the platoon can pose a danger to following vehicles.
- Minimizing jerks: Depending on the speed and actions taken by vehicles in the platoon, a vehicle can experience rapid switching between acceleration and deceleration, which can cause passenger discomfort. To ensure passenger comfort, jerk should be kept within approximately 0.1–0.2g/s (where g is the acceleration due to gravity).

- Optimizing gaps to maximize vehicle throughput and safety: The time headway may need to be adjusted based on speed and traffic conditions.
- Adapting to "cut-in" and "cut-out" vehicles: Vehicles may abruptly cut into the gaps in a platoon or abruptly leave a platoon. The vehicles in the platoon have to be able to respond to such situations without compromising safety or passenger comfort.
- Platoon formation and admission control: Well defined procedures and communication protocols are needed for advertising the existence of a platoon, obtaining relevant information about a platoon, and joining and leaving a platoon.

The controller in the CACC model described above uses information from only the immediately preceding vehicle. This can be extended to use the information from the lead vehicle of the platoon [19, 20] and other vehicles in the platoon, to improve performance.

All of these requirements suggest a need for a more robust communication system than the currently deployed cellular technologies, in order to support efficient and large scale platooning. The following main features of 5G NR (New Radio) enable it to overcome the limitations of LTE and support platooning applications more efficiently.

- Cell capacity and spectral efficiency: NR supports much wider bandwidths than LTE (up to 400 MHz). Additionally, multiple-input multiple-output enhancements lead to at least a doubling of spectral efficiency even in sub-6 GHz deployments. Use of LDPC coding for the data channels also improves spectral efficiency. As a result, NR can support significantly more vehicles per cell site.
- Reliability: NR has been designed to be able to support ultra-reliable communication. In particular, the NR control channel uses Polar codes to improve reliability of the control channels. NR also supports various other techniques to improve reliability. These include repetitions of transmissions, improved CSI reporting and packet duplication at the PDCP layer. Studies indicate that, with these enhancements, NR can achieve reliability of 99.999% or higher [21].
- Latency: While the CACC performance is not very sensitive to latency (e.g. in the analyses above, a latency of 100 ms was assumed), various other aspects of platooning benefit from low latency. For example, lower latency communication helps in cases of abrupt braking and cut in of vehicles described above. Furthermore, lower latency also makes it easier to achieve platoon stability. NR achieves a latency of 4 ms (one way, between the device and the base station). This is further reduced to 1 ms if mini-slots are used.

In addition to the above, mobile edge computing is a key feature of 5G. It not only enables a reduction of delays in the communication path for applications but also enables other applications which rely on localized information. These aspects are discussed next.

9.4.2 Edge Computing for Platooning

While the CACC approach described above forms the core of platooning, there are several other features needed for successful and widespread platoon operation. Examples of such features include being able to handle vehicles of different types (e.g. a large vehicle following a smaller vehicle), adapting to weather and traffic conditions, managing merging and exiting vehicles, etc. The edge computing capabilities of 5G can be leveraged to support such features and significantly improve platoon operation.

Roadside base stations (Roadside Units [RSUs]) can provide 5G connectivity to vehicles. All intelligence related to platooning can reside in a platooning server that is connected to the RSUs, as shown in Figure 9.10. Vehicles communicate relevant information to the platooning server, which then acts on the received information and transmits commands to the vehicles.

The platooning server can enable functions well beyond CACC, as described below:

1) Core platooning functions: Vehicles periodically report their speed, target acceleration, and observed inter-vehicle gap with respect to the preceding vehicle. For each vehicle, the platoon server uses information from the

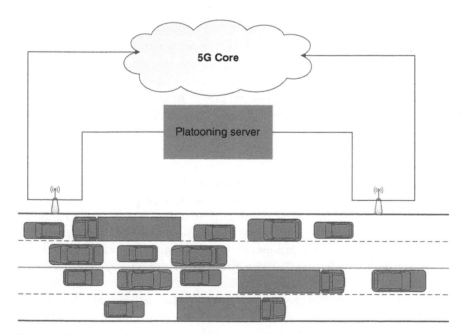

Figure 9.10 Edge computing for platoons.

preceding vehicle and the platoon leader to determine a command (i.e. an adjustment to acceleration/deceleration) and communicates this command to the vehicle. The platoon server can also utilize information from other intermediate vehicles in addition to the lead vehicle and the preceding vehicle. It is noteworthy that the platoon server can maintain two control loops for each vehicle – one with respect to the preceding vehicle and the other with respect to the lead vehicle – and transmit a command to the vehicle based on the "safer" of the two. A vehicle performing such control by itself would not be able to do this since it would not have an estimate of the distance to the lead vehicle (sensing is limited to the preceding vehicle). The platoon server on the other hand collects information from all vehicles and is able to estimate, for each vehicle, the gap with respect to the lead vehicle.

2) Platooning functions customized to vehicles: Larger and heavier vehicles generally require longer durations to accelerate to a target speed or to brake. This impacts the safety of heterogeneous platoons. This can be handled by the platooning server by customizing the platoon control and commands for each vehicle. For example, for larger vehicles a larger value of engine time lag τ may be used. Similarly, for larger vehicles the time headway to the preceding vehicle can be larger. Vehicles report the relevant information such as weight, number of axles, etc. before the platoon establishment. In addition, vehicle diagnostic information can be made available to the platoon server to account for situations where some essential functions may not be functioning optimally.

3) Taking into account location specific information: Although highways are designed for uniformity, there are significant variations in terrain and topography even over a few miles. Knowledge of the topography can improve the platooning performance by adjusting target speeds and accelerations. Additionally knowledge of upcoming curves in the roadway can help assist lateral control. The information about the topography and the roadway can easily be pre-provisioned into the server with location coordinates. It is impractical to pre-provision such information in a vehicle, since it is not possible to know in advance where the vehicle is going to travel.

4) Taking into account traffic/incident information and weather information: Traffic information is generally available today through crowdsourced platforms and can provide advance knowledge of traffic incidents and congestion. Incorporating such information into the platoon server enables better routing and management of the platoon. For example, in heavy traffic, the platoon may be broken up into smaller platoons or entirely disbanded. Weather conditions can affect performance of vehicles in different ways. Knowledge of weather conditions can enable adaptation of platooning procedures to ensure safer travel. In addition, real-time information from vehicles, such as wheel slippage, can be used to guide following vehicles.

5) Managing of vehicles entering and leaving a platoon: Vehicles leaving or join-ing a platoon can disrupt the platoon structure and if the disruption is not quickly controlled, the platoon can disintegrate. While a vehicle leaving or joining the platoon is recognized by the vehicle immediately following (via its sensors), relying on the sensors alone can result in dramatic and uncomforta-ble changes to speed. Instead, a platoon server can create a large enough gap for a vehicle joining the platoon to smoothly join the platoon and smoothly close the gap when a vehicle leaves the platoon.

6) Platoon establishment and dissolving: A well-established protocol for estab-lishing and dissolving a platoon is an essential requirement for platoon opera-tions. Such a protocol has to not only establish the sequence of vehicles but also needs to have means to authenticate vehicles. A rogue driver/vehicle that is able to join or lead a platoon can cause serious damage. The security and authentication capabilities that cellular networks are known for can be used to ensure that rogue actors cannot take control of platoons.

9.5 High Definition Maps Use Case

An HD map is a machine-readable map which allows an automated vehicle to know where it is and what is around it. This type of map comprises various tiled layers that are highly accurate (centimeter level accuracy). Each tile contains a section of a map that covers a given geographical area. Utilizing the concept of tiles has several advantages because it allows the procedures related to map updates to be optimized. Instead of the vehicle having to download entire maps, individual tiles of interest can be provided to the vehicle, allowing for optimiza-tion in terms of latency and bandwidth requirements during map downloading and map updating procedures. The tiles of interest for a given vehicle usually depend on the vehicle location and the vehicle route. If a route is known or can be predicted, then only the tiles in and around the route need to be provided to the vehicle. Moreover, when changes occur in the environment (such as new roads added), only the tiles of interest are affected and only those need to be updated in the vehicle.

A typical configuration with the main components that may be involved in this use case are depicted in Figure 9.11.

The components shown in the figure contain three different applications of the HD maps:

1) HD map client application: The HD map client application is installed in the vehicle. This application is responsible for maintaining the most up-to-date map of the surroundings. It uses stored information (e.g. a reference map),

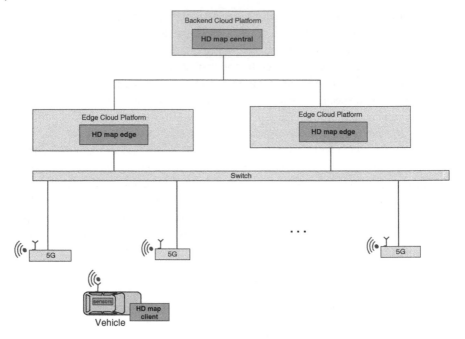

Figure 9.11 Generic HD maps use case main components.

information collected by the vehicle sensors, and information received from other vehicles and/or the network in order to update the HD map.

2) HD map edge application: The HD map edge application is installed in the edge cloud platform, which is also referred to as the MEC platform, which was discussed in Section 9.3. This application handles localized information and is responsible for updating the information in vehicles located in a given geographical area. The geographical area of interest may be predefined and fixed, or it may dynamically change. The HD map edge application communicates directly with the HD map client and HD map central applications.

3) HD map central application: The HD map central application is responsible for the large scale aggregation of information and modification of large scale maps. This application is aware of all of the HD map edge applications running in the different edge cloud platforms, and their respective geographical area of interest. This application is responsible for configuring the relationship between the HD map edge applications and their geographical area of interest.

As previously discussed, by placing content in the MEC platform, the user can access the content much faster, and without increasing the bandwidth requirements toward the backend cloud. As a result, more users can be supported in the network. In addition, applications can have access not only to local content but also to real-time information related to the network conditions.

9.5.1 HD Maps Procedures

Typical procedures involved with HD maps are map downloads and map updates. Here we provide an overview of how these procedures are usually implemented and how they are usually partitioned between the vehicle, the edge platform, and the backend (cloud) platform.

9.5.1.1 Map Download Procedure

When the vehicle first requests a map of a given area, it determines its current location and then sends the request to the HD map edge application at the edge cloud platform, with the current location included in the request. Requests for maps also include the time stamp of when the map was last updated, so the network only needs to send the changes since the last update. Based on the location, the HD map edge application determines the tiles that need to be provided. The total number of tiles to be provided is usually determined by the HD map edge application at the edge cloud platform. Typically, a few tiles corresponding to the area around the vehicle need to be provided. If the vehicle route is known and provided to the edge application, then tiles in and around the route may also be provided to the vehicle.

If the edge cloud server does not have all tiles of interest, it may request them from the HD map central in the backend cloud platform. Once the tiles of interest are received from the backend cloud server, they are provided by the edge cloud server to the vehicle. As the vehicle moves, new tiles may be requested.

The map download procedure is depicted in Figure 9.12.

Figure 9.13 shows a typical data flow for the HD map download procedure in more detail. The Sensor Data Processing function receives information gathered by the sensors and processes it in order to deliver processed information to other functions. This function also determines the vehicle location based on the sensor information. With the location information available, the Vehicle Tile Handler function checks for maps that are stored and determines if tiles corresponding to the location are available. If the required tiles are not available, a request for the tiles is sent to the Vehicle Request Handler function to forward the request to the edge.

The Edge Request Handler function receives the request and forwards to the Security Manager function, which authenticates the client and authorizes the transaction. The Edge Tile Handler will then process the request. If a request for map tiles is received, it uses the location and time stamp provided in the

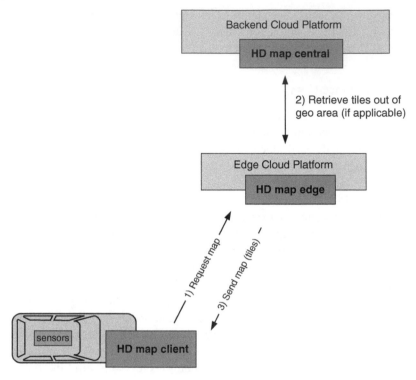

Figure 9.12 HD map download procedure.

request and determines which tiles to send. If the tiles are available locally, then it sends the tiles to the client application. If the tiles are not available locally, it sends a request for the tiles to the HD map central application at the backend cloud platform.

At the backend cloud server, the Cloud Tile Handler uses the location and time stamp provided in the request to determine which tiles to send.

9.5.1.2 Map Update Procedure

The other procedure of interest is the one where a change in the surroundings is reported by vehicle and the map is updated. In this process, a sensor-equipped vehicle compares the map that was acquired from the edge cloud server to its own observation of the environment (via sensors) and determines deviations with significant size. For example, a vehicle may observe via its sensors the presence of a roadside barrier that is not present in the HD map currently stored. The deviations are then sent to the HD map edge application at the edge cloud platform. The HD map edge application then evaluates and integrates this information. The independent observation (from independent vehicles) of this deviating region enables

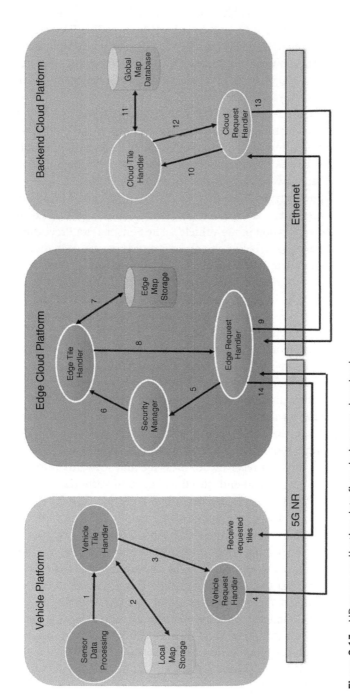

Figure 9.13 HD maps application data flow during map download.

the edge cloud server to increase the confidence and accuracy of the spatial change on the map. Once such a spatial change has been reflected into the local map stored in the edge cloud server, then a propagation to the backend server shall follow in a regular update time slot. Once new information is provided to the backend server, the HD map central application is responsible for distributing the new information to a relevant area.

The update map procedure is illustrated in Figure 9.14, where vehicle 1 observes the changes in the environment via its sensors, and reports the differences to the HD map edge application. Note that in this example vehicle 2 is not equipped with the same sensors but it still learns about the changes in the environment via information exchanged between the vehicles (through the network).

Figure 9.15 shows a typical data flow for the HD map update procedure in more detail. A reference map is assumed available in the vehicle (i.e. previously uploaded to the application in the vehicle). The Sensor Data Processing function receives information gathered by the sensors and processes it in order to deliver processed information to other functions. The captured and processed data is provided to the Map Comparator and Aggregator function, which compares the stored environment information with that captured by Sensor Data Processing. When a map deviation is detected, this function aggregates the map deviations with existing tiles to make a new updated reference map.

The Vehicle Tile Handler function is continuously performing tile acquisition as the vehicle moves and is the one that decides when the map deviation needs to be sent to the network for an update. This decision may be based on the dynamic nature of the information: the vehicle may be configured to only report static/structural changes, such as new roads and buildings, as opposed to for example reporting the presence of a road construction barrier, which is temporary.

Once the vehicle sends the information to the edge cloud server, the Edge Request Handler function receives the message and forwards it to the Security Manager function, which authenticates the client and authorizes the transaction. The Agent Consensus Manager decides if the detection of the deviating region gives enough confidence. These updates are usually decided based on crowd-sourcing, which means that the Agent Consensus Manager may wait for several reports of the same deviating region from different vehicles in order to establish consensus. If consensus is reached, then the deviation is sent to the HD map central application in order to update the global database. The edge cloud server may also send the deviation to other vehicles in the region it is serving.

The Cloud Tile Handler function receives the updated map/tile and decides if the changes are to be incorporated and if so, modifies and stores the updated tiles. The Cloud Tile Handler function also decides if the information needs to be sent to other edge cloud servers, based on the tile location and geographical responsibility of each HD map edge.

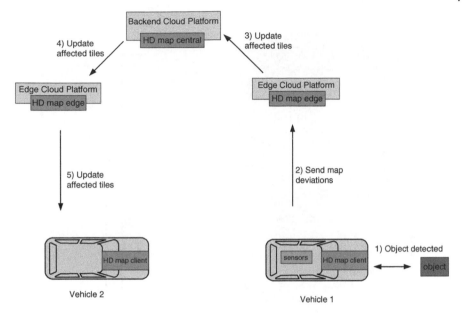

Figure 9.14 HD map update procedure.

9.5.2 Edge Computing for HD Maps

As discussed in the previous sections, by having a differentiated and hierarchical level of responsibility between the edge platform and the backend platform, it is possible to optimize the latency and the bandwidth utilization of the system, maximizing the network capacity. Each edge cloud platform is responsible for a predetermined geographical area. The geographical area may be chosen based on the number of vehicles that need to be supported in the region, the latency requirements for the services supported, and the traffic generated by the users in the given region. A large geographical area minimizes the number of edge servers needed but may impact latency and capacity. A small geographical area provides the optimal latency and capacity, at the cost of more edge servers.

Another benefit of this model is the ability to scale the number of edge servers needed as the number of users to be serviced increases in any given region or as more services get deployed. Initial deployments may assume a single edge platform covering a very large area. As autonomous driving becomes more popular, more servers can be deployed in the same region. This flexibility reduces the initial cost of deployment.

RSUs can be utilized for communication between vehicles and edge servers. RSUs provide wireless connectivity between vehicles and computing devices deployed at the side of the road. Usually these computing devices are targeted

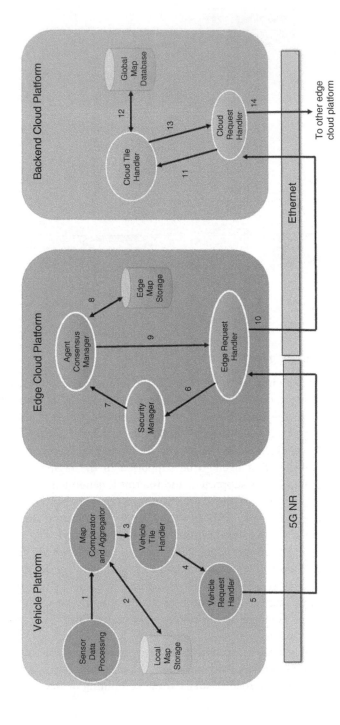

Figure 9.15 HD maps application data flow during map updates.

toward road traffic optimization. HD maps functionality can be seen as one of such services. It is important to note that the deployment of RSUs is not likely to be ubiquitous and, when not present, base stations (or gNBs) can be utilized for the communication between the vehicle and the edge server.

Another advantage of utilization of edge computing capabilities for the HD maps use case is that it allows for location services to be incorporated in maps. Traffic and weather information can be displayed in the map to help the driver or can facilitate optimal route selection. Advertisements can be incorporated into the maps, especially if a route is known (e.g. sending coupons for stores that are on the user's route).

The use case for HD maps can be extended to a real-time situational awareness (RTSA) map. As compared with HD maps, which is the use case to distribute information on relatively slow changing road conditions, a RTSA map distributes real-time information to traffic participants, including vehicles, bicycles, and pedestrians. Use cases such as a pedestrian crossing a street and bicycle warnings are examples of RTSA map updates. These are notifications sent from RSUs to vehicles regarding the presence of pedestrians crossing the street or bicycle ahead warnings. Such notifications require very stringent requirements on latency and reliability. The URLLC (ultra-reliable and low-latency communications) features of 5G-NR, combined with edge computing, can be used to facilitate such use cases. These use cases would utilize a procedure similar to the map update procedure described in the previous section but without the need of crowdsourcing information. In the RTSA map use case, the Agent Consensus Manager decides on the deviating confidence based on a single report (either from a vehicle or from a local video near an intersection). A single report in this case would be sufficient to reach consensus and other vehicles in the same geographical area can be notified of the presence of a pedestrian or bicycle. In this case, because the changes are temporary, the system may not update the global database. This is yet another advantage of using a MEC platform for this use case: the temporary changes stay confined to the local platform, without impacting the central database. The updates are done in a fast and efficient manner, thus meeting the service requirements without wasting communication, compute and memory resources.

9.6 Summary

The next decade is likely to see major changes in how we operate vehicles and how transportation systems function. Several of these changes are already underway and the general trend of these changes is the automation of various functions in vehicles and the reduction of human involvement in driving. Many of these changes are driven by technological advances in sensing, robotics and applied machine learning.

Sensors have fundamental limitations that relate to range and line-of-sight. Radar, lidar, and cameras cannot see past obstructions. Communication has played and will continue to play an important role in making vehicular automation more effective.

One of the key features of 5G is MEC. MEC enables placement of computing and storage resources closer to physical locations where they are needed. Given that much of the information related to driving is geographically local, MEC can provide significant benefits to connected vehicle use cases. It not only enables a reduction in application latency but it also allows for deployment of targeted infrastructure and computing resources specific to the connected vehicle use cases. For example, map information or image recognition capabilities can be placed near roads and can serve vehicles and RSUs.

While we have not discussed all possible use cases in the connected vehicle space, we cover two exemplary use cases. Platooning of vehicles is widely seen as the first step toward significant automation of vehicles on highways, particularly trucks. It has significant benefits in terms of improving safety and reducing driver fatigue, and it also provides significant fuel savings. Platooning is based on CACC which requires a vehicle to receive information from vehicles ahead of it and perform corresponding adjustments to its speed and acceleration. In particular, receiving information periodically about the engine action applied by the preceding vehicle or vehicles enables nearly perfect tracking of vehicles even in very long platoons. If the communication between vehicles is error prone (i.e. messages are lost or delayed), the vehicle occupants can experience discomfort in the form of jerk (which increases with larger proportions of lost messages). The higher reliability of 5G-NR communication can greatly reduce the possibility of such jerk. Additionally, MEC can enable platooning features that go well beyond the longitudinal control of vehicles and are essential to platoon operation. Examples of such features include support for vehicles cutting-in and cutting-out, customized control for different types and sizes of vehicles, and adaptation based on location-specific information.

HD maps are essential for higher levels of automation of vehicles. HD maps are machine readable maps that enable a vehicle to locate itself within and navigate through its surroundings. They provide information about surroundings at a very fine granularity. Up-to-date maps are critical for automated driving. Collection of information for HD maps is done via crowdsourcing and can be supplemented with sensors placed in the vehicle and at RSUs and reported to an edge server. Significant computational and storage capabilities are needed to keep the HD maps up to date. In particular, it is necessary to validate the collected information and aggregate information collected from multiple sources. The information then has to be disseminated to vehicles. The MEC architecture combined with the 5G-NR low-latency communication can achieve the timely updates of the maps to vehicles.

Acknowledgment

The HD maps work described in this chapter was greatly assisted by the expertise of Dr.-Ing. David Gonzalez-Aguirre.

References

1 Commission for Global Road Safety (2006). Make Roads Safe. A New Priority for Sustainable Development.

2 Dalal, K., Lin, Z., Gifford, M., and Svanstrom, L. (2013). Economics of global burden of road traffic injuries and their relationship with health system variables. *International Journal of Preventive Medicine* 4 (12): 1442–1450.

3 The Economist (2014). The cost of traffic jams http://www.economist.com/the-economist-explains/2014/11/03/the-cost-of-traffic-jams (accessed 25 January 2019).

4 Schneider, B. Traffic's mind-boggling economic toll http://www.citylab.com/transportation/2018/02/traffics-mind-boggling-economic-toll/552488 (accessed 25 January 2019).

5 Arnott, R. and Small, K. (1994). The economics of traffic congestion. *American Scientist* 82 (5): 446–455.

6 Critical Reasons for Crashes Investigated in the National Motor Vehicle Crash Causation Survey, DOT HS 812 115, February 2015.

7 CAR 2 CAR Communication Consortium (2017). Road Safety and Road Efficiency Spectrum Needs in the 5.9 GHz for C-ITS and Automation Applications, February 2017.

8 5G Automotive Association (2016). The Case for Cellular V2X for Safety and Cooperative Driving, November 2016.

9 Preliminary Statement of Policy Concerning Automated Vehicles, May 2013.

10 MEC in 5G networks, ETSI White Paper No. 28, June 2018. https://www.etsi.org/images/files/ETSIWhitePapers/etsi_wp28_mec_in_5G_FINAL.pdf (accessed June 2019).

11 https://www.etsi.org/technologies/multi-access-edge-computing (accessed June 2019).

12 Mobile Edge Computing (MEC); End to End Mobility Aspects Group Report, ETSI GR MEC 018 V1.1.1 (2017–10).

13 Draft, Multi-access Edge Computing (MEC); MEC Application Mobility Draft, ETSI GS MEC 0021 V2.0.8 (2019–04).

14 Ioannou, P. and Chien, C.C. (1993). Autonomous intelligent cruise control. *IEEE Transactions on Vehicular Technology* 42 (4): 657–672.

15 Shladover, S.E. (1991). Longitudinal control of automotive vehicles in close formation platoons. *Journal of Dynamic Systems, Measurement, and Control* 113 (2): 231–241.

16 Tsugawa, S. (2014). Results and issues of an automated truck platoon within the energy ITS project *Proceedings of the. IEEE Intelligent Vehicles Symposium (IV)* (June 2014).

17 Volvo Trucks successfully demonstrates on-highway truck platooning in California https://www.volvotrucks.us/news-and-stories/press-releases/2017/march/volvo-trucks-successfully-demonstrates-on-highway-truck-platooning-in-california (accessed 25 January 2019).

18 Milanes, V., Shladover, S.E., Spring, J. et al. (2014). Cooperative adaptive cruise control in real traffic situations. *IEEE Transactions on Intelligent Transportation Systems* 15 (1): 296–305.

19 Ploeg, J., Scheepers, B.T.M., van Nunen, E. et al. (2011). Design and experimental evaluation of cooperative adaptive cruise control *14th International IEEE Conference on Intelligent Transportation Systems*, Washington, DC, USA (5–7 October 2011).

20 Narasimha, M., Desai, V., Calcev, G. et al. (2017). Performance analysis of vehicle platooning using a cellular network *Proceedings of the. IEEE Vehicular Technology Conference* (September 2017).

21 3GPP TR 37.910 (2019). Study on Self Evaluation Towards IMT-2020 Submission.

10

Connected Factory

Amanda Xiang and Anthony C.K. Soong

Futurewei Technologies, Plano, TX, USA

Abstract

This chapter first discusses the capabilities of 5G as defined in Release 15 that can be used for the manufacturing industry and describes the use cases needed for the manufacturing industry. The illustrated use cases can be roughly classified into five major characteristics: factory automation, processes automation, human–machine interfaces and production IT, logistics and warehousing, as well as monitoring and maintenance. It should be clear that a particular use case may have one or more of the characteristics. The chapter discusses each in order. In Release 16, 3GPP will start to close the gaps from the information and communication technology 5G to that needed by the smart factory. As 3GPP completes its work, it can be readily seen that all aspects of the untethered communication can be met by new 3GPP releases and will make 5G a key enabler for redefining the manufacturing process.

Keywords *5G technologies; factory automation; humanmachine interfaces; logistics; manufacturing industry; predictive maintenance; process monitoring; processes automation; use cases;warehousing*

10.1   Introduction

The commercial 5G wireless service offers unprecedented capacity, reliability, connectivity, and coverage. As a result, it has also garnered unprecedented interest among many different vertical industries. Future smart factories will be among the first to employ 5G to transform their industry. The vision is that every part of the

5G Verticals: Customizing Applications, Technologies and Deployment Techniques, First Edition.
Edited by Rath Vannithamby and Anthony C.K. Soong.
© 2020 John Wiley & Sons Ltd. Published 2020 by John Wiley & Sons Ltd.

factory is fluid except for the floors, walls, and ceilings. Every other part of the factory, machines, devices, and vehicles will be mobile and enabled by 5G. It is, thus, not surprising that the majority of these use cases are to transition the communication technology in interconnecting sensors, actuators, and controllers for automation from wired bound to untethered wireless technology. Untethering the communication not only allows for the reconfiguration of a manufacturing line to be significantly lower, but it also enables the future smart factory to be flexible and scalable. It allows the factory owner to change production lines according to demand.

10.2   5G Technologies for the Manufacturing Industry

Before we discuss the use cases needed for the manufacturing industry, it is instructive to understand the current capabilities of 5G as defined in Release 15 that can be used for the manufacturing industry.

10.2.1   Ultra-Reliable and Low-Latency Communications

For Release 15, the ultra-reliable and low-latency communications (URLLC) service has the capability to achieve an end-to-end latency of around 1 ms with a reliability of 99.999% [1]. This is already good enough to support most of the closed loop control applications but that of the most extreme use cases such as, for example, motion control (see Section 10.4.1). Notwithstanding, the capability is certainly good enough to begin the transition to smart factories. Release 16 will further enhance the support for the smart factory and the first release specifically designed with factory support.

For mobility, [2] evaluated the URLLC service via outdoor trial and confirmed that the 3GPP requirements can be satisfied with 25 km/h mobility. Further enhancement will be needed in future releases to cover that use case.

10.2.2   Enhanced Mobile Broadband

The enhanced mobile broadband (eMBB) service defined in Release 15 can achieve downlink average capacity of 16.65 bit/s/Hz/transmit receive point with massive MIMO (multiple-input multiple-output) of 32 transmit antennas and 4 received antennas [3]. For 100 MHz channel with only one transmit and receive point (no multi-user MIMO), the average data rate is 1.665 Gbits/s. For the uplink, with massive MIMO of 4 transmit antennas and 32 received antennas, the average capacity is 6.136 bits/s/Hz/transmit receive point. For example, if the user equipment (UE) only uses 20 MHz of the uplink channel then the average data rate is 120 Mbits/s. This type of data rate is sufficient for almost all but the most demanding use case, such as high-end cameras (see Section 10.4.13).

Mobility support for many manufacturing services is not much of a concern except for motion control and mobile robots which demand high mobility support. The 5G eMBB service was designed from the beginning to support mobility. It has robust mobility mechanisms capable of supporting seamless mobility across the network.

5G also already has the capability to support traffic classification and prioritization natively. The eMBB services were designed from the beginning to have a flexible quality of service (QoS) framework that supports a variety of traffic flows with a range of QoS requirements.

Unlike the traditional wireless network which has been built as a monolithic network, 5G, from the beginning, has been designed to provide flexible support for different services through the network slicing feature. Loosely speaking, a network slice corresponds to a dedicated set of virtual or physical resource on the same physical network infrastructure and dedicated behavior. A more technical description of the network slice can be found in [3] but this loose definition is sufficient for our discussion here. A particular application can be served by one or more network slices; this is dependent on implementation. For our discussion here we can limit it to a single slice. The slice provides the isolation needed for the applications to scale without impacting other applications on the network thus guaranteeing the Service Level Agreement (SLA) of a particular service no matter how other services (slices) are scaled. This also means that regardless of what is happening in the other slices, certain level of security of a particular slice is maintained. From a particular application point of view, it is as if it has dedicated network resources, even though the underlying implementation is not dedicated. This concept of isolation is key to understanding the paradigm shift features that 5G brings to the table.

What it means for manufacturing use cases is that the different use cases may be implemented on their own dedicated slice. Thus, it naturally resolves the conflicting requirement of different manufacturing use cases. For example, the robotic use case in a plant can scale completely independent of the closed loop control use case. It allows, for example, different lines of communication to scale without impacting the others. For a business operating multiple plants, it will even let the communications of the different plants to scale dynamically and independently. All of this is available natively in the Release 15 eMBB services.

10.2.3 Massive Machine Type Communication

The 5G system defined in Release 15 can easily exceed the massive machine type communication (mMTC) requirements from the International Telecommunication Union (ITU). For narrowband Internet of Things (NB-IoT), it will support around 600 000 connections per square kilometer per 180 kHz of bandwidth [3]. That means that for 1 MHz of bandwidth it can support over 3.3 million connections. This number is sufficient for all manufacturing scenarios. Since the number of connections is not really seen as a limiting factor, Release 16 development is focused on enhancing the downlink and uplink efficiency, as well as providing extreme coverage [4].

The battery life for NB-IoT devices was calculated in [5] to be of the order of 10 years if transmission of the device is kept to 200 bytes per day with a coupling loss of −164 dB. This feature is especially important for sensors that need to be untethered such as, for example, some of those used in the logistic use case. If long-term battery life is an issue, energy harvesting can be used.

These release 15 performance capability are sufficient for the initial senor implementations of the factory. But more capability enhancements, such as security, relay capability, data rate and latency, are needed to meet those more advanced sensor implementations in the future smart factory.

10.3   5G Alliance for Connected Industries and Automation

The 5G Alliance for Connected Industries and Automation (5G-ACIA) was formed in order to have the industrial vertical requirements adequately addressed by the mobile communication industry (Figure 10.1). Its aim is to be "the central and global forum for addressing, discussion, and evaluating relevant technical, regulatory, and business aspect with respect to the 5G domain" [6]. Its members include operation technology (OT) companies as well as information and communication technology (ICT) companies. It was formed in April 2018 and provides, among other things, the forum for OT and ICT players to come together to formulate a set of unified requirements from the entire ecosystem for 3GPP to develop features needed by the manufacturing industry.

At present, there are 56 member companies including Siemens, Bosch, Vodafone, T-Mobile, Orange, China Mobile, Ericsson, Huawei, Nokia, Intel, and Qualcomm. The activities of 5G-ACIA are structured into five working groups (Figure 10.2):

- WG1 collects and develops the requirement and use cases. It provides those use cases to 3GPP SA1 as well as to educate the ecosystem on the existing 3GPP requirements. It serves not only to align the industry but to also identify gaps in the 5G system to be addressed for the need of the manufacturing industry. The liaison between 5G-ACIA WG1 and 3GPP SA1 is eased because many members belong to both groups.
- WG2 identifies and articulates the spectrum needs of the manufacturing industry and explores new operator models. It also coordinates 5G-ACIA activities with relevant regulatory bodies.
- WG3 considers the overall architecture of the future 5G-enabled industrial network. It also investigates possible integration concepts and migration paths,

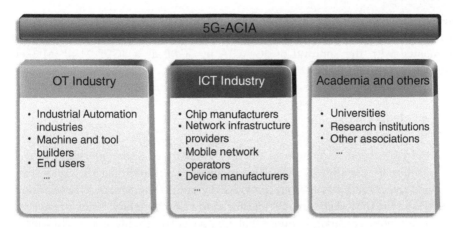

Figure 10.1 A pictorial representation of the 5G-ACIA ecosystem.

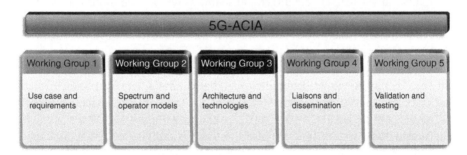

Figure 10.2 The structure of 5G-ACIA.

as well as evaluating key technologies from 5G standard bodies. It has completed a radio propagation analysis and evaluation of the factory environment which was key in starting the corresponding work in 3GPP. It is working on studying the usages of private networks as well as the seamless integration of 5G and industrial Ethernet technologies.

- WG4 is responsible for the interaction with other initiatives and organizations. It does this by establishing liaisons and initiating suitable promotional activities.
- WG5 deals with the final validation of 5G for industrial applications, which includes the initiation of interoperability tests, larger trials, and potentially dedicated certification procedures.

3GPP has recognized the significance of vertical industries in the success of 5G and is working to build and foster relationships with all vertical industries interested in 5G. 5G-ACIA can be viewed as an upstream forum for 3GPP. It provides 3GPP with new use cases and requirements. The role of 3GPP is to standardize features in 5G to support these use cases. Based upon these standards, 5G-ACIA can then evaluate the gaps and validate the use cases. This close relationship is cemented with the approval of 5G-ACIA by 3GPP as a 3GPP MRP (Market Representative Partner).

10.4 Use Cases

5G-ACIA has been studying how the future factory can take advantage of the newly offered capabilities of 5G [7]. It concluded that the defined 5G services can provide the capabilities that are provided by both Time-Sensitive Networking (TSN) and the Industrial Ethernet for the factory. Consequently, 5G may converge the many communication solutions that are used currently. Moreover, as a widely accepted commercial wireless communication system, it can, for the first time in the industrial vertical, enable direct and seamless wireless communication from the field level to the cloud. In this respect, it will enable the flexibly and versatility needed to improve the productivity of smart factories and achieve the vision set forth in Industrie 4.0.

A number of use cases for the "Factory of the Future" have been identified where 5G can make a significant impact. A diagrammatic representation of these use cases and their requirements in terms of the three major service features of 5G, namely eMBB, URLLC, and mMTC, are shown in Figure 10.3. The axis can be interpreted, loosely, as URLLC stands for reliability and latency, eMBB for capacity, wide area coverage and mobility, and mMTC for massive connectivity and energy efficiency. It can be seen that no use case exists that requires the full functionality of all three major services nor even full features of two major services and that the use cases are either eMBB, mMTC, or URLLC centric. Furthermore, it can be argued that since the first phase of 5G has quite good eMBB features, URLLC feature development is the more important feature to focus development effort on for this vertical. It should be noted that as 5G becomes more mainstream in the smart factory, the capability may need to grow. For example, as more and more sensors are used in the smart factory, the use cases may grow in the mMTC direction.

The illustrated use cases in Figure 10.3 can be roughly classified into five major characteristics [5]: Factory automation, Processes automation, HMIs (human–machine interfaces) and production IT, logistics and warehousing, as well as monitoring and maintenance. It should be clear that a particular use case may have one or more of the characteristics. We will now discuss each in order:

- **Factory automation** has to do with the automatic control, monitoring, and optimization of factory processes and workflows. It is regarded by many as the key enabler of economical high-quality mass production. Since quality and low

mMTC

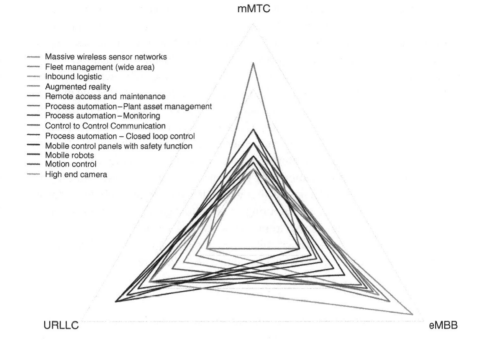

— Massive wireless sensor networks
— Fleet management (wide area)
— Inbound logistic
— Augmented reality
— Remote access and maintenance
— Process automation – Plant asset management
— Process automation – Monitoring
— Control to Control Communication
— Process automation – Closed loop control
— Mobile control panels with safety function
— Mobile robots
— Motion control
— High end camera

URLLC

eMBB

Figure 10.3 A representation of the various use cases for the industrial vertical.

cost are paramount here, some of the use cases in this class require the highest reliability and lowest latency. This is especially critical as the factory of the future moves from static lines to a novel modular production system offering high flexibility and versatility.

- **Process automation** has to do with the automation of the controls and handling of substances, such as food, chemicals, liquids, etc., in the production facility. A typical closed loop manufacturing process, in general, consists of sensors that are measuring process values, such as temperature, pressure, distance, etc. that provide information for the controller to control a manufacturing system or subsystem. The manufacturing process facility of interest here can be highly variable depending on the manufacturing process. For example, its size can range from 100 to 1000 m^2. Indeed, the process may even run over multiple geographically disjointed facilities.
- **HMIs and production IT** has to do with the HMIs of production devices, such as panels on a machine, IT devices, such as computers, laptops, printers, etc., as well as IT-based manufacturing applications, such as a manufacturing execution system (MES) and an enterprise resource planning (ERP) system. On top of the currently existing human interfaces, the factory of the future may make use of new interfaces using virtual reality (VR) and augmented reality (AR)

technologies. Both the MES, which monitors and documents how raw materials are transformed into the finished product, and the ERP system, which provides a continuously integrated view of the important business processes, require large data from the production process be available in a timely fashion.

- **Logistics and warehousing** involve the storages and flows of materials and products in industrial manufacturing. Logistics is defined as the control of the management of goods and information flow and may be intra-logistics, which deals with logistics in one site, or inter-logistics, which deals with logistics between different sites. Warehousing is the storage of materials and products. In this class of use cases, the location, tracking and monitoring of assets are of prime importance.

- **Monitoring and maintenance** refer to the monitoring of processes and/or processes outside of a control loop. In other words, it is monitoring without having an immediate impact on the manufacturing process. It ranges from condition monitoring and predictive maintenance to machine learning for future improvements to the manufacturing process. As a result, the consistency of the data is of paramount importance and the latency in which the data is delivered is relatively unimportant. As more and more sensors will be deployed in the factory of the future, the size of the data may be large even though the data from each sensor may be small.

Having a good understanding of the characteristics, we will discuss the details of some of the use cases in the following. As we go through the discussion, we will see that in addition to the usual 5G performance requirements, these use cases generally also contain operational and functional requirements. Examples of operational requirements include simple system configuration, operation, management, and SLA assurance mechanism. Examples of functional requirements include: security, functional safety, authentication, and identity management. It should now be clear that some of these requirements are similar to that needed for a commercial wireless system. This should not be surprising given the excitement about 5G for the industrial vertical is partly due to commercial 5G's synergy with industrial vertical needs.

10.4.1 Motion Control

Motion control in an industrial manufacturing sense is defined as the control of moving parts in a machine in a well-defined manner. The nature of the motion may be linear, circular, or even more complex. Wireless connection is envisioned here to replace the wired connection because of the freedom of movement that it allows and the possibility of reduction in down time from the wear and tear movement makes on the wired connectors.

A motion control feedback loop in diagrammatic form is shown in Figure 10.4. As usual, the input to the loop is the desire output. The controller, based upon the input from the sensors and the input to the loop, periodically sends a control signal to the actuator. The actuator then performs a corresponding action on one or more

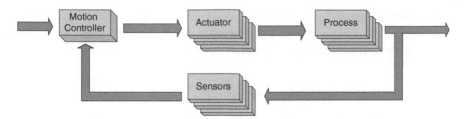

Figure 10.4   The motion control feedback loop.

processes, which in this case, by definition, is a movement of a part in the process. At the same time, the sensors monitor the process output and report the results to the controller. All of this is done in a strictly deterministic cyclic manner. The current industrial Ethernet solution supports a period of around 50 µs. It is now clear that smaller periods allow for faster and more accurate movement controls. Also, the reliability of the communication may impact both safety and the quality of the output. This leads us to requirements that will highly emphasize the URLLC service that could be beyond the URLLC capability in Release 15. 3GPP RAN will study this and determine how best to support such strict URLLC requirement; for example, whether to use the U_u or the side link (PC 5[1]).

Besides the transmission of time critical data in motion control, some additional non-real time (NRT) data is exchanged between the nodes in the motion control loop. This data may be transmitted in parallel to the real time data discussed previously. Examples of this type of data are firmware/software updates or maintenance information. The capacity requirement of these NRT data is not large and data rates of around 1 Mits/s should be sufficient and can be easily handled with the current capabilities of the eMBB feature.

It is also within reason that, in a smart factory, not all devices will be connected wirelessly in the motion control system. Given that, it is then necessary for the 5G system to coexist with the industrial Ethernet system. This implies the following:

1) 5G network should be able to forward frames from the Ethernet sources to 5G destinations and vice versa.
2) Precise time synchronization between multiple motion controllers, some of which are connected to industrial Ethernet and some to 5G can be solved if both industrial Ethernet and 5G supports the Precise Time Protocol (IEEE 1588).
3) As the Ethernet device can be separated from other devices on the same physical Ethernet network via Virtual LAN (IEEE 802.1Q), it means that the 5G system must now become aware of Virtual LAN associations.

---

1 This is the current terminology in 3GPP. However, there is as of yet no agreement on NR terminology within the vehicle-to-everything work item, and so as the development of Release 16 progresses further the terminology may change for the NR side link.

4) In order to overcome communication bottle necks, industrial Ethernet solutions use reservation protocols. This implies that 5G would need to be aware of the time aware scheduling defined in IEEE 802.1Qbv.

10.4.2 Control to Control Communication

Control-to-control (C2C) communication, which is the communication between controllers, is already in use today. There are two major use cases:

1) Large machines (such as newspaper printers) use several controllers to cluster machine functions together. For efficient operation, these controllers need to communicate among themselves. As can be readily seen, this type of communication uses real time communication that needs to be synchronized.
2) Communication between the controllers of individual machines is often necessary for the efficient operation of the factory. An example of this is the communication between machines that coordinate and control the handoff of work products.

It is not too difficult to imagine that in the factory of the future, with the untethering from 5G, such communication will only increase in both connection and the amount of data exchanged. From a requirements point of view, the main focus can be the C2C communication between different motion subsystems because it typically has the most demanding requirements. In general, we can say that the stringent requirements of reliability, latency, and determinism are similar to that for motion control except that the service area can be much bigger; up to 100 times.

10.4.3 Mobile Control Panels with Safety Functions

Control panels are used to provide the human interface to production machinery and interaction with moving devices. Their main function is to provide configuration, monitoring, debugging, control, maintenance, and emergency safety stop of industrial machinery ranging from simple machines, robots, and cranes to entire production lines. From the safety point of view, reliability and speed are the critical requirements. The URLLC requirements are similar to that of motion control but there are additional requirements on eMBB because the service area is usually bigger although interaction with the public cellular network is not necessarily required.

It is also clear that not all aspects of control panel communication deal with safety. Consequently, it also has the requirement for simultaneous transmission of non-critical data with the critical safety data. For this non-critical requirement, the 5G eMBB feature is more than sufficient for the task.

10.4.4 Mobile Robots

Robots are programmable machines that perform multiple operations. They and autonomous machinery, such as automated guided vehicles, have gained more and more usage in the manufacturing environment. Therefore, it is expected that they will play an ever-increasing role in the factory of the future. They have functionality in a localized service area, such as inside a factory, as well as in a wide service area, such as outdoors. For the localized service area, ultra-low latency is required but connectivity to the public network is not. For the wide service area case, the latency requirement can be relaxed but interconnection with the public network (e.g. service continuity and roaming) is additionally required. As robotic functions become more complicated, there will arise a need for synchronized ultra-low latency and ultra-high reliable communication between robots, for example to control and coordinate carrying and handoff of work goods, or to assemble products in highly cooperative and automatically way. However, not all communications are latency sensitive and so there is also the requirement of parallel transmission of non-critical data.

10.4.5 Massive Wireless Sensor Networks

The factory of the future will be no different than other 5G environments. Sensors, which measure or monitor the state or behavior of a particular environment, will become ubiquitous. The types of sensors and their communication needs will be heterogeneous. The environments in which the sensors operate will be dynamic, ranging from very benign to very hostile. They may function individually or be used in a coordinated distributed system. Independent of how the sensors are deployed, the training and analysis of the sensor network may be accomplished in a centralized fashion, in a distributed fashion, or a mixture of both.

The traffic pattern of sensors is also heterogeneous and depends upon the type of sensor and the environment in which it operates. It may need low bandwidth as well as high bandwidth. The reliability ranges from strict for safety sensors to that of NRT traffic.[2] The traffic may exhibit cell similarity and/or periodicity. To minimize the load on the communication system, pre-processing of the data may be employed.

The architecture of the 5G network that supports sensor networks may also be different. In some cases, such as in remote areas it may be best to have them connected to large macro cells while in other areas it may be better to take advantage of the heterogeneous network architecture.

---

2 Reliability here implies packet reliability. For NRT traffic, reliability of the data can be maintained, for example using Hybrid Automatic Repeat Request or Automatic Repeat Request even though the transmission of each packet itself is unreliable.

It now becomes obvious that massive sensor networks, such as the Internet of Things (IoT), do not represent a single use case. They rather represent a class of very diverse use cases. Although the current LTE based mMTC can meet many of those requirements, there are limitations to restrict them to meet the future factory and process plants. New 5G capabilities are needed, such as low latency, high data rate, wider coverage and suitability for non-public network deployment.

10.4.6   Remote Access and Maintenance

Remote access and maintenance will be a key paradigm shift for the factory of the future. Clearly untethering the factory would transform what can be controlled and maintained remotely, as well as what constitutes "remotely." We can see that the freedom from the telephone wire that mobile users enjoy so much will directly translate into the smart factory environment.

Remote access and maintenance will apply to devices which:

1) Already have a communication connection, which could be cyclic, for transmitting data regularly.
2) Act almost autonomously.
3) Have local interaction with humans.
4) Sleep most of the time and are only woken up to establish the connection for transmission.

For case 1, the ad hoc communication for remote access and maintenance would be in parallel with the regular data communication. For case 2, the device must have its own local processing power, since it functions almost autonomously, and only establishes communication when necessary. In case 3, if the local personnel need further assistance, a remote connection can be established so that they can access remote expert assistance. Case 4 is similar to a sensor that only wakes up once in a while to transmit its information.

What constitutes remote need not mean extreme geographic differences. Consider, for example, where the partner is a mobile device. It may use remote access even when it is very near the device; maybe even geographically collocated. From the device point of view, it would not and have no need to differentiate that it was accessed locally. From the remote partner's point of view, however, it may be of benefit for it to know what devices are local, say, within the same plant. This could potentially be achieved in the application layer if the application also had access to the 5G location services.

Tracking of the inventory of the devices and periodic readouts of configuration data, event logs, revision data, and predictive maintenance information is also another interesting use case. The requirements in this use case can easily be met with the 5G eMBB service.

Depending on the action instructed by the remote partner, the action of the remotely accessed device may have significant impact on the functionality of the factory. With that being said, the mere act of remote access should not impact the functionality of the factory around the device being accessed, i.e. the act of just opening a remote connection should have no impact on the factory functionality.

The ability to severely impact the factory operation by the remote partner brings up the need for cyber security. Most of the current wired communications do not consider cyber security because the physical access to the device and network will restrict the access to authorized personnel. Thus, we will need a suite of security protocols to protect the device against cyber threats, ranging from turning remote access off to different restrictions on who can access the device and what they can do with the device.

The last thing to consider in this use case is the life cycle of industrial machines. Industrial devices are expected to have a life cycle of around 25 years which is longer than the typical life cycle of a generation of wireless technology (10 years). This means that either we need a way to easily upgrade the communication of the device or that we will need to consider that older generations of wireless technology will need to coexist with newer generations. This is not necessary a standards issue but mainly an implementation issue.[3] Nevertheless, the standards will need to take this into account when designing the system.

10.4.7 Augmented Reality

AR will be a major game changer, like everywhere else, on how we will interact with our environment. On the factory floor of the future smart factory, workers will be optimally supported in their new tasks and activities, as well as in ensuring smooth operation. The following applications are foreseen:

1) Monitoring processes and production flows.
2) Expert assistance, maybe even step-by-step instructions, for a specific task.
3) Ad hoc support from remote experts.
4) Training and education.

As comfort and ergonomics will be of prime importance in AR devices for the workplace, energy efficient communication and processing comes to the forefront. One solution is to make use of the edge computing offered in 5G to offload all complex processing tasks to the edge.

Another key requirement is that the augmented image should track with the movement of the worker. This puts some limitation on the location of the edge processing as well as the latency of the transmission. Consequently, this use case, similar to motion control, has strict requirements for latency. The required service

---

3 There are a number of ways this could be handled in implementation. For example, the device's communication subsystem could be engineered to be upgradable over the air using the remote access and maintenance use case.

area, however, is bigger than that for motion control. It does not require interaction with the public network for local service but remote expert support may need the support of the remote access and maintenance use case depending upon the level of support.

10.4.8 Process Automation – Closed Loop Control

This use case consists of the case where multiple sensors are installed in a plant, the sensors send their measurements to the controller, usually periodically, and the controller then decides on how to control the actuator to give the desired output of the process. This use case is different to motion control in the sense that the controller is not controlling moving parts. Consequently, it has stringent requirements for latency but not as extreme as that in motion control. The service area is bigger than that for motion control but interaction with the public network is usually not required.

10.4.9 Process Automation – Process Monitoring

This use case consists of multiple sensors installed in the plant whose main purpose is not to control the plant in the short term but rather to give insights into the process, environmental conditions, or inventory of material. The key requirement of this use case is wide area coverage and interaction with the public network may be required. The current 5G eMBB service seems to be sufficient for this use case. In the case of an extreme number of sensors, augmentation with, e.g. NB-IoT, would solve the connection issues [3].

10.4.10 Process Automation – Plant Asset Management

For a factory to run optimally, each part of the factory must also be running optimally. This use case involves using sensors to monitor the performance of plant assets, such as pumps, valves, heaters, etc. and to maintain them in a timely fashion. Most of the requirements of this use case can be met by the eMBB service.

For the monitoring of processes, environmental conditions inventory of material, and asset management maintenance, the positioning requirement of the sensors used shows some similarity to that for IoT devices; for indoors it should be better than 1 m, 99% of the time.

10.4.11 Inbound Logistics

Logistics deals with the organization and management of things, including non-physical things such as information between different points, e.g. between points of origin and points of consumption. It includes all aspects of transportation and storage of things as they move between the points. Inbound logistics has to do with the logistics of things coming into a business. The major use case here includes the tracking of goods, and transportation assets. For example, heavy goods vehicles can be connected to the public 5G network to enable real time tracking and telematics. The container or pallets of goods can be connected to the public 5G network for tracking and inventory control purposes. Both of these use cases can be supported by the current eMBB service. The pallet or container may also detect the presence of a local 5G non-public network as it moves into the intended plant. This implies that it must know which non-public network it should connect to; for example, in consolidated trucking, as a truck moves into a business, those goods intended for that business should be able to connect to the non-public network. However, the non-public network should not be able to have any knowledge of other goods on the truck; ideally, they would not even attempt to attach to the non-public network. In this case there are several options as the 5G or IoT device attached to the pallet detects the presence of the local 5G non-public network:

1) Dual connection: The 5G or IoT device has an independent subscription to both the 5G public network and the local 5G non-public network.
2) Dual connection: The 5G or IoT devices remains connected and registers on the public 5G network and establishes a simultaneous parallel connection to the local 5G non-public network.
3) Manual PLMN selection: The 5G or IoT device performs a manual PLMN selection procedure, which may be initiated automatically. This requires that the local network has a human readable identifier to enable manual selection.

The moving of goods and materials inside a plant is also of considerable interest. In this use case, these goods and materials are wirelessly connected and tracked as they move through the plant. In this case the major requirements are seamless service continuity between 5G public and 5G non public networks and indoor positioning accuracy that needs to be less than 1 m (maybe in the range of 20–30 cm) because the communication needs are rather benign.

10.4.12 Wide Area Connectivity for Fleet Maintenance

In this use case, the interest is in using 5G for automatic wide area data collection and tuning of an automotive fleet. Besides the usual use case for telemetry, the 5G capability allows for the downloading of electronic control unit (ECU) tuning to the heavy goods vehicle in a semi static manner, for example to optimize the performance based upon the load, telemetries, and environmental conditions. The major requirement here is the wide area coverage capability because the data requirements are benign. The data itself can be quite delay tolerant; maybe even longer than 30 min because the ECU tune will not be changed that frequently. The reliability of individual uplink telemetries is not so critical as long as periodicity of the data is frequent enough; it is not that big a deal if you miss an individual message.

10.4.13 High-End Camera

Recently there has been discussion among a number of manufacturing companies about a use case where high-end cameras are connected via 5G on the factory floor for quality control and inspection within the production line. Another use is for mobile robots to send back what the robot is seeing. By high-end camera we are talking about high resolution with around 100 frames per second. It then becomes clear that the uplink data rate and coverage in a highly complex radio environment will be the key requirements. The density of theses cameras for production is of the order of 10–20 cameras in an area of 100 m by 50 m initially and it may grow from there.

10.5 3GPP Support

Since the OT network of a smart factory is very different to that of a traditional IT network, with different deployment models and requirements, its requirements can be very challenging to an ICT-oriented mobile network, which at its heart is an IP-based IT network. 3GPP, as the standard organization that developed 3G, 4G and 5G technologies has taken up the challenge to ensure that 5G technology can be deployed into the future smart factory from Release 16. The Release 15 standard is the first standalone 5G NR standard (published in June 2018) providing the basic capabilities of 5G and is eMBB application centric. Release 16 is expected to be completed around March 2020, and may be considered by many vertical players as the first 5G vertical-ready standard release for initial commercial deployment.

The *Industrie 4.0* and smart factory vertical applications bring very different and challenging requirements for 5G. This required 3GPP to be willing to accept new concepts and ideas in order to develop innovative solutions to address future smart factory needs for 5G. Fortunately, because smart factory is a very promising, massive market for 5G, many key 3GPP ICT companies such as Huawei, Ericsson, Nokia, Vodafone, and other mobile operators, are working closely with smart factory OT partners, such as Siemens, Bosch, 5G ACIA and its members, to provide strong and collaborative 3GPP support in various 3GPP working groups:

- A working group (SA1) has defined a clear 5G use case and requirements for release 16 and 17 to guide the downstream working groups.
- Technical downstream working groups (RAN1, RAN2, SA2, SA3, SA5, SA6) have started to develop suitable 5G standard solutions based on the requirements from smart factory applications.

10.5.1 5G Use Case and Requirements for Smart Factory

In order to develop the right 5G standard solution, clear 5G service requirements from smart factories based on the right use cases are needed. This is the work of 3GPP SA1 WG. In March 2017, Siemens successfully created a study item (FS-CAV) in SA1 WG on communication for automation in vertical industries, which initiated the 3GPP effort to develop suitable 5G for *Industrie 4.0*. With the creation of 5G ACIA in April 2018, more voices from smart factory OT players are being heard. With strong collaboration between ICT and OT companies in SA1 WG as well as 5G ACIA, FS-CAV was completed, with its study report (23.208) published in May 2019. This report became an extremely important specification for 3GPP Release 16 and had a significant influence on the other 3GPP working groups. It has led to many new technical solutions, such as URLLC enhancement, TSN integration, new private network, and so on. Later, based on this study report, SA1 developed a standard specification (TS22.104) which was completed in November 2018 to define the first set of smart factory specific normative 3GPP service requirements for release 16 including all the QoS key performance indicator (KPI) requirements, for which the 3GPP downstream working groups will develop technical solutions with the successful completion of release 16 TS22.104. Furthermore, with the strong support of the OT and ICT industries, SA1, on November 2019, completed its release 17 TS22.104 enhancement with new requirements.

In OT systems, service performance is measured based on end-to-end system performance which includes the application layer, while 3GPP can only address the performance in the network transport layer. Therefore, a clear boundary for 5G requirements is needed. The communication service interface (CSIF) is introduced as the border between application and 5G communication service. Source and destination CSIFs are used as two performance measurement points for 3GPP to define 5G network performance, as illustrated in Figure 10.5.

There are many different applications associated with the different traffic types which may have different QoS KPI requirements. 3GPP SA1 groups the different QoS requirements, and use cases into four traffic classes for factory and process automation applications. Each traffic class has a different set of performance characteristic parameters as well as influence parameters which are not essential

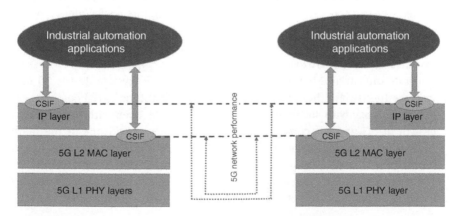

Figure 10.5 5G network performance measurement in an OT network.

for the performance of an item but do affect its performance, such as service area, number of users, and survival time:

- Periodic deterministic communication: This is periodic with stringent requirements on timeliness and availability of the communication service.
- Aperiodic deterministic communication: This is without a pre-set sending time but still with stringent requirements on timeliness and availability of the communication service.
- Non-deterministic communication: This includes periodic/aperiodic non-real-time traffic and subsumes traffic types other than periodic/aperiodic deterministic communication.
- Mixed traffic: This is traffic which cannot be assigned to one of the above communication patterns exclusively.

SA1 has defined new 5G requirements in some key areas for smart factory in Release 16:

- Network QoS performance requirements. It defined much more stringent KPI requirements, especially on latency and service reliability and availability, such as for some periodic deterministic traffic, such as motion control applications with one way communication latency as low as $500\,\mu s$, with communication service availability of 99.999–99.999 99%, and communication service reliability (mean time between failures) of about 10 years.

  However, some system performance measurements used by traditional industry OT systems are different to that of the mobile network. Therefore, how to translate the OT's QoS KPIs to ICT's network KPIs is sometimes difficult. One example is the communication service availability defined above, which cannot be mapped to existing 3GPP defined network KPIs straightforwardly. From the OT perspective, the communication service availability requirement is the combination of latency, survival time, and reliability requirement for the 5G system, and the system is considered unavailable to the cyber physical application when an expected message is not received (e.g. transfer time is bigger than the maximum transfer latency) by the application after the application's survival time expires. There is ongoing effort in 3GPP SA1 and 5G ACIA to define a relatively clear mapping of KPIs between the two systems, and this is expected to be achieved in the coming new 3GPP release.

- Integrate Industrial Ethernet, such as TSN, into the 5G system. In the OT system, the Industrial Ethernet currently is the dominant industrial communication technology with 46% market share (Figure 10.6). Even with the projected growth of 5G wireless communications, it will still remain one of the main communication technologies (especially TSN) to link components together in the factory in the near future.

  Therefore, the 5G system must support the seamless integration and interplay with the Industrial Ethernet. Since TSN, defined by 802.1Q, is considered as the

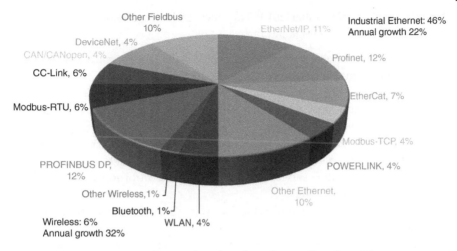

Figure 10.6　2017 Industrial network market share. Source: Data from [8].

main technology replacing existing Ethernet and Fieldbus technologies deployed in factories for many motion control applications, 3GPP Release 16 is focused on supporting TSN as defined in IEEE 802.1Q in the 5G system. The TSN 5G integration requirements are not only about how to support TSN traffic as one type of PDU session but also support clock synchronization defined by IEEE 802.1AS across 5G-based Ethernet links and other Ethernet transport such as wired Ethernet networks and Ethernet passive optical networks (EPONs).

- Support new private networks for smart factory. Because of the stringent QoS performance requirements, data privacy, security considerations, and other special characteristics of OT network operation, there is a strong need from smart factory players to deploy 5G factory networks in a private network format which are not influenced by other public users. The private network can be deployed by mobile operators with some shared resources with public networks, such as using network slicing or network sharing technologies, or be deployed as standalone networks managed by either the factory owner or mobile operators with their own security mechanisms and dedicated network resources. 3GPP SA1 defined the non-public network (NPN) for this kind of vertical private network in order to distinguish it from the public network (PLMN) normally used in 3GPP. Because the NPN will be deployed differently to the public network, some new requirements on network selection, private user authentication and authorization and interworking between the NPN and the public network are defined.

3GPP SA1 has completed the first set of service requirements for smart factories in 3GPP Release 16, and just completed the additional new service requirements for release 17 on November 2019. The new service requirements which were introduced in release 17 for smart factories are:

- Industrial Ethernet integration, which includes time synchronization, different time domains, integration scenarios, and support for TSN.
- NPNs as private slices, and further implications on security for NPNs.
- Network operation and maintenance in 5G NPNs for cyber-physical control applications in vertical domains.
- Enhanced QoS monitoring, communication service and network diagnostics.
- CSIF between application and 5G systems, e.g. information to network for setting up communication services for cyber-physical control applications and the corresponding monitoring.
- Network performance requirements for cyber-physical control applications in vertical domains.
- Positioning enhancements for Industrial IoT, including relative positioning information and vertical directions/dimension.
- Device-to-device/ProSe communication for cyber-physical applications in vertical domains.

10.5.2 5G Standardized Solution Development for Smart Factory

After 3GPP SA1 WG has defined the 5G service requirements for smart factory for Release 16, the other 3GPP working groups have developed or are developing the technical solutions from different system component perspectives to fulfill those service requirements in Release 16, some working groups also start the work for release 17 since July 2019.

10.5.2.1 5G System Architecture for Smart Factory (SA2)

3GPP SA2 WG, which is the working group that is identifying the main functions and entities of the network from a 5G end-to-end system perspective, created a work item (Vertical_LAN) for potential 5G enhanced solutions to support Industrial IoT for Release 16 in 2018, and is getting close to completing the normative specification work based. Additionally, 3GPP SA2 created several new study items for release 17 which are related to the new requirements from 3GPP SA1, such as:

1) FS_IIoT: Study on enhanced support of industrial IoT - TSC/URLLC enhancements;
2) FS_eNPN: Study on enhanced support of Non-Public Networks.

For Release 16, SA2 WG has developed some system architecture enhancements to address some key requirements from SA1:

1) New UE network selection mechanism for selecting the NPN. Since the traditional 3GPP-based mobile network is mainly mobile operator oriented, the UE network selection standard solutions are mostly about selecting the mobile

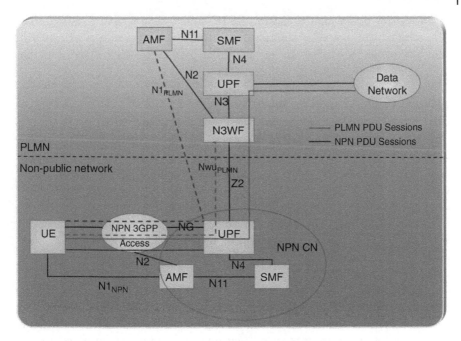

Figure 10.7 The service continuity architecture and solution.

operator's public network first. This is not suitable for the smart factory's private network deployment, e.g. for standalone private network deployment, in which the devices in the factory may only care about the network ID of the private network. Therefore, SA2 defined a new non-public network ID (NPN-ID) to identify a NPN for standalone deployment. This ID will be used along with public land mobile network ID (PLMN ID) consisting of mobile country code (MCC) 999 (assigned by ITU for private networks) and an mobile network code (MNC) defined by 3GPP to identify the cell as part of a NPN.

2) In order to allow some users within a NPN network to be able to access the applications in the public network, a service continuity architecture and solution was developed (Figure 10.7[4]). This solution mostly leverages the existing non-3GPP (e.g. WiFi) access solution to reduce overall system impact considering 3GPP already supports some enterprise networks using WiFi to access the 3GPP network. The solution uses over-the-top IP security tunneling between UE and the N3IWF function in public networks to protect the NPN and the public network. This solution with N3IWF (Non-3GPP InterWorking Function) also allows UE to access NPN via a public network.

- For TSN integration support, in order to reduce 5G system impact and for easy brownfield deployment, SA2 chose solutions which use the 5G system

---

4 A list of the network function acronyms is given at the end of the chapter.

as a TSN bridge, and supports TSN Ethernet as one type of data in a PDU session. The proposed solution is illustrated In Figures 10.8 and 10.9 in which the TSN translator can be either part of the User Plane Function (UPF) (Figure 10.8) or outside the UPF (Figure 10.9).

10.5.2.2 Other 3GPP Work for Smart Factory

One of most important requirements from the smart factory for 5G deployment is that the 5G radio must provide URLLC service. 5G-NR is able to deliver 1 ms latency for Release 16, with further enhancement in capacity of URLLC support, further reduced latency and improved coverage expected for Release 17. The other important work in RAN WG for the smart factory is to provide high accuracy and flexibly deployed clock synchronization solutions for the factory devices, because the OT system, especially systems with motion control (Section 10.4.1), heavily rely on highly accurate clocks for synchronization within a working domain as well as crossing multiple working domains which may belong to one or different base stations.

Security is not only a key feature of 5G but is also a high priority requirement from smart factory for 5G deployment. SA3 WG (System Architecture WG 3) is the working group responsible for security and privacy in 3GPP systems. The working group determines the security and privacy requirements, and specifies the security architectures and protocols. They are looking into 5G security features to support vertical applications by also creating a study item on security for 5G's enhanced support of vertical and LAN services for Release 16. Most of their work focuses on security for verticals' private network. One of the challenges and a controversial issue in Security WG which is still open is whether to introduce other alternative security credential and mechanisms which have not been defined by 3GPP but are being used in existing OT systems. Because existing 3GPP security mechanisms, such as Extensible Authentication Protocol and Authentication and Key Agreement (EAP-AKA) and Subscriber Identity Module (SIM), are the foundations of the existing 3GPP systems and global mobile operator's core operation, any changes can have not only significant technical impact on the current 5G architecture but significant impact on the business model being widely used today by ICT players. Therefore, any changes need to be fully studied in order to reach agreement within the industry.

3GPP SA5 WG is the working group responsible for network operation and management. Since the private network (NPN) can become one of the main deployment models for the verticals, especially for smart factory, and the private network is a much simpler and smaller scale network compared with the massive public network, the network Operation, Administration and Management (OAM) system for 5G needs to be enhanced and optimized for private network deployment. SA5 WG has created a new study item in this area, which will lead to normative 5G OAM system features for smart factory's private network in Release 16.

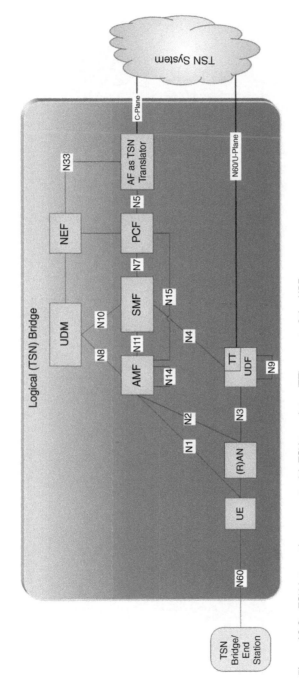

Figure 10.8 TSN integration support with TSN translator (TT) as part of the UPF.

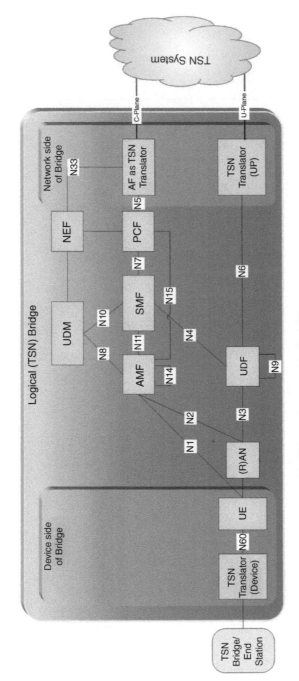

Figure 10.9 TSN integration support with TSN translator outside the UPF.

10.6   Early Deployments

10.6.1   Spectrum

One of the keys to any smart factory deployment will be the spectrum. In the early days of this concept, there was a thinking that this could be deployed in unlicensed spectrum. As the industry gained more understanding of the use case, it became clear that the unlicensed spectrum cannot guarantee the reliability needed for OT networks (see the discussion in Section 10.4) and that a certain amount of licensed spectrum is needed. The location of this licensed spectrum, however, needs careful consideration. It must be considered and prioritized a globally harmonized 5G band for the mobile commercial service. Currently, the 3400–3600 MHz band, within the C band (3300–4200 and 4400–5000 MHz), is allocated to mobile services on a co-primary basis in almost all countries throughout the world. The 5G NR specification will support 3300–3800 MHz from the beginning using a test-driven development access scheme. This band is in line with the allocation plans from many countries and is the ideal spectrum for the harmonized 5G band. Thus, it is recommended that countries allocate at least 100 MHz of contiguous bandwidth from this band to each commercial 5G network [9]. This implies that countries should allocate the private network spectrum for the OT outside of this band or if they want to put some of the OT private spectrum in this band, care should be taken to leave sufficient spectrum for the usage of the commercial public network.

Perhaps the country that is furthest along in allocating spectrum for OT services is Germany. The current status is that the 3400–3800 MHz band will be divided; 300 MHz (3400–3700 MHz) for the public network and 100 MHz (3700–3800 MHz) for the NPN. The OT industry can bid for the non-public part for the usage of the smart factory. The bidding for the public 5G spectrum, which also includes spectrum in the 2100 MHz band, started in March 2019 and concluded around June 2019. The regulations (including fees) for the non-public part have been finalized currently. The application procedure has officially started since November 2019.

Although 100 MHz spectrum for OT applications seems like a large amount of spectrum, if the vision of all the use cases for the smart factory becomes a reality, it is likely that more than 100 MHz will be needed. Fortunately, for many of the use cases, the coverage is small, compared with the commercial macro cell, and that means it can take advantage of higher frequency bands. The World Radiocommunication Conference 2015 (WRC 15) identified several frequencies for study within the 24.25–86 GHz range (the higher frequency bands identified at WRC 15; Figure 10.10). Groups 30 and 40 are prioritized and all regions and countries are recommended to study and support these two bands for IMT, as well as aim

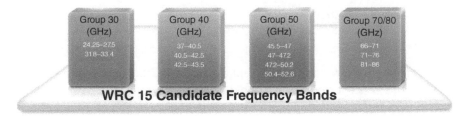

Figure 10.10 The higher frequency bands identified at WRC 15.

to harmonize the technical conditions for the support of these bands at WRC 19. The OT industry should then study the usage of these bands for smart factories.

10.6.2 Early Trials

There have been a number of announced projects and demonstrations of the smart factory use case in the last few years. This section will give a brief summary of these activities. The list should not be considered comprehensive but is intended to give the reader a sense of the tremendous activity and enthusiasm within the manufacturing vertical for 5G.

Among the first cooperative efforts to speed up the development of smart factory is the cooperation between Huawei and Toshiba announced in 2017. This project is to collaborate on the integration of NB-IoT for the development of smart factories. The goal is to accelerate the commercial availably of NB-IoT in a diverse range of deployment use cases applied to real-world manufacturing scenarios. Initial testing will be performed in Huawei's NB-IoT Open Lab located in Shanghai. The second phase will be live field tests to develop a suite of smart factory solutions.

In the 2018 Hanover Messe, Bosch demonstrated an extended smart factory mock-up that included a Nokia 5G network enabling real-time data exchange and artificial intelligence fault pre-emption.

In April 2018, Nokia Telia and Intel announced a smart factory test using a video application to monitor and analyze a process on the factory floor at the Nokia factory in Oulu, Finland. The application incudes a machine learning algorithm to alert the operator of inconsistencies in the process. A second test showed that real time data can be rendered and accessed at Telia's data center that is 600 km away in Helsinki.

Huawei and Beckhoff Automation GmbH & Co. demonstrated the wireless communication between two cooperating programmable logic controllers during the 2018 Hannover Messe. This proof of concept showed that with direct integration of 5G into the industrial controllers, industrial automation can be realized in a more flexible and economical manner. The two companies also verified predictive maintenance with a wireless sensor scenario.

Ericsson reported in 2018 that in a case study that applied cellular IoT technology to its Nanjing Ericsson Panda factory breakeven can be reach in less than two years with 50% return on investment in the first year [10]. The use cases investigated include asset monitoring, environmental sensors, and inbound logistics. The study showed both CAPEX and OPEX savings.

Qualcomm and Bosch announced in February 2019 the completion of a joint research effort to study the radio channel for industrial environment and characterize its features. Using this as a springboard, the two companies intend to launch an over-the-air trial using 5G NR industrial IoT. Among other things, the trial will investigate the usage of 5G Release 15 NR for current industrial usage.

In February 2019, the UK turned on its first ever 5G factory trial. The trial is being conducted by the Worcestershire 5G Consortium and used the Worcester Bosch factory as the test bed. The organizations involved in the trial include AWTG, Huawei, O_2, Malvern Hills Science Park, British Telecom, Worcestershire Local Enterprise Partnership, 5GIC at the University of Surrey, QinetiQ, Yamazaki Mazak, and Worcester Bosch. It will investigate methodologies to increase industrial productivity via "preventative and assisted maintenance using robotics, big data analytics and AR over 5G" [11]. Besides investigating the applicability of 5G for the manufacturing industry by taking the usual measurement of speed and latency, the Consortium will endeavor to make this trial network ultra-secure with the help of 5G.

10.7 Conclusions

Factory operations have come full circle starting from the first industrial age with craft production, moving to mass production during the global age, and now moving toward hyper-customization. Consumers are no longer satisfied with mass produced goods but are pushing for customized goods produced just for them; the so-called lot size of one manufacturing. To enable such flexibility in the factory, digitization and automation will be key technologies along with untethered communication. This is the promise of the smart factory that will realize the vision of *Industrie 4.0*. The OT industry and the ICT industry have come together in a number of fora to ensure that 5G can provide for the untethered communication of the smart factory. A number of use cases have been identified and their requirements compared with the defined 5G services of eMBB, URLLC, and mMTC, have been studied. In Release 16, 3GPP will start to close the gaps from the ICT 5G to that needed by the smart factory. As 3GPP completes its work, it can be readily seen that all aspects of the untethered communication can be met by new 3GPP releases and will make 5G a key enabler for redefining the manufacturing process.

Acronyms

The network function acronyms as given by [12]:

AF   Application Function
AMF   Core Access and Mobility Management Function
AUSF   Authentication Server Function
DN   Data network, e.g. operator services, Internet access or third party services
NEF   Network Exposure Function
NRF   Network Repository Function
PCF   Policy Control Function
(R)AN   (Radio) access network
SDSF   Structured Data Storage Function
SMF   Session Management Function
UDM   Unified Data Management
UDSF   Unstructured Data Storage Function
UE   User equipment
UPF   User Plane Function

References

1 3GPP TR45.804 (2015). Cellular system support for ultra-low complexity and low throughput Internet of Things, v.13.1.0. https://portal.3gpp.org/desktopmodules/Specifications/SpecificationDetails.aspx?specificationId=2719.

2 Iwabuchi, M., Benjebbour, A., Kishiyama, Y. et al. (2018). Evaluation of coverage and mobility for URLLC via outdoor experimental trials. *2018 IEEE 87th Vehicular Technology Conference (VTC Spring)*, Porto.

3 Lei, W., Soong, A.C.K., Jianghua, L. et al. (2020). *5G System Design: An End to End Perspective*. Cham: Springer International Publishing.

4 Ericsson (2019). RP-190770 Revised WID: Additional MTC enhancement for LTE. https://www.3gpp.org/ftp/tsg_ran/TSG_RAN/TSGR_83/Docs.

5 3GPP TR22.804 (2018). Study on Communication for Automation in Vertical domains (CAV), v.16.2.0. https://portal.3gpp.org/desktopmodules/Specifications/SpecificationDetails.aspx?specificationId=3187.

6 5G-ACIA (2019). About 5G-ACIA. https://www.5g-acia.org/index.php?id=5036 (accessed 12 April 2019).

7 5G-ACIA (2019). 5G for Connected Industries and Automation (White Paper - Second Edition). https://www.5g-acia.org/index.php?id=5125.

8 Carlsson, T. (2017). Industrial Ethernet and Wireless are growing fast – Industrial network market shares 2017 according to HMS. https://www.hms-networks.com/press/2017/02/20/industrial-ethernet-and-wireless-are-growing-fast-industrial-network-market-shares-2017-according-to-hms (accessed 18 April 2019).

9 Huawei (2017). 5G Spectrum: Public Policy Position. http://www-file.huawei.com/-/media/CORPORATE/PDF/public-policy/public_policy_position_5g_spectrum.pdf.

10 Ericsson (2018). The world's first cellular IoT-based smart factory. https://www.ericsson.com/assets/local/networks/offerings/celluar-iot/connected-factory-case-study.pdf.

11 UK 5G Innovation Network (2018). Worcestershire 5G Consortium. https://uk5g.org/discover/testbeds-and-trials/worcestershire-5g-consortium (accessed 16 April 2019).

12 3GPP TS 23.501 (2019). System architecture for the 5G system (5GS), v.16.0.2. https://portal.3gpp.org/desktopmodules/Specifications/SpecificationDetails.aspx?specificationId=3144.

Index

5G Verticals: Customizing Applications, Technologies and Deployment Techniques, First Edition.
Edited by Rath Vannithamby and Anthony C.K. Soong.
© 2020 John Wiley & Sons Ltd. Published 2020 by John Wiley & Sons Ltd.